THE 7 STEPS TO
EMERGENCY
PREPAREDNESS
FOR FAMILIES

Funded by a California
State Library, Crisis
Collections grant 17/18

THE 7 STEPS TO
EMERGENCY
PREPAREDNESS
FOR FAMILIES

A Practical and Easy-to-Follow Guide to Prepare for Any Disaster

Kim Fournier CD, MA

Library and Archives Canada Cataloguing in Publication Data

The 7 steps to emergency preparedness for families / Kim Fournier.

Includes index.

ISBN 978-0-9950360-0-0

1. Emergency management--Handbooks, manuals, etc.
2. Survival--Handbooks, manuals, etc.
3. Safety education.
4. Preparedness.

HV551.2.F68 2016   363.34'7   C2016-905454-3

Printed in Canada

Layout and cover design by Cara Robbins

First Paperback Edition

Dedicated to our children's resilience and survival.

Table of Contents

Seven Steps to Preparedness ix

Preface xi

Part I: Get Ready 1

Step 1: Safeguard Important Documents 3

Step 2: Create a Communication and Reunion Plan 5

Step 3: Create a Shelter and Evacuation Plan..13

    Shelter In Place 13

    Evacuate..13

Step 4: Assemble Emergency Supplies 19

    Emergency Supplies 19

    Grab-and-Go Bag...20

    Disaster Supplies for the House..25

    Food and Water 26

    Work, School, and Vehicle Kits 31

    Preparing Your Pet for a Disaster 31

Step 5: Prepare the House...35

    General House Preparation Guidelines...35

    Utilities...40

    Multi-Residential Complexes 45

Annual Maintenance 48

Part II: Respond 49

Step 6: Respond to a Disaster 51

    Shelter in Place or Evacuate 51

    Shelter in Place...52

    Evacuate Safely..53

    Living in a Mass Care Shelter...54

    Stay Informed During a Disaster 55

    Power Outages 56

    Response Actions for Unique Hazards..57

Part III: Return Home..65

Step 7: Return Home Safely..67

    General Considerations..67

    Flood Safety and Clean Up 68

Summary 73

Additional Information..75

Disaster 101 77

    The Disaster Management Process 77

    Common Hazards 79

Hygiene and Sanitation...83
    Basic Practices...83
    Handling Garbage and Other Waste...84
    Hygiene and Sanitation Checklist..86
Collecting Water from Your Hot Water Tank...87
Collecting and Disinfecting Water..89
    Disinfection Methods...89
    Filtration Methods...90
    Sedimentation..90
    Distillation...90
    Collecting and Treating Water from Alternate Sources....................91
Disaster Mental Wellness...93
    Psychosocial Responses to Disasters..93
    Reducing the Emotional Effects of a Disaster..................................94
Special Needs...97
    Planning for Specific Requirements..97
Disaster Insurance...101
    General Insurance Tips..101
    Insurance Checklist...102
Shopping List...105
Acknowledgements...111
About the Author...113
Products and Services...114

Seven Steps to Preparedness
An all-hazards approach

Get READY

Step 1: Safeguard Important Documents (page 3)

Step 2: Create a Communication and Reunion Plan (page 5)

Step 3: Create a Shelter and Evacuation Plan (page 13)

Step 4: Assemble Emergency Supplies (page 19)

- ✔ a grab-and-go bag
- ✔ stockpile of items for the home
- ✔ work, school, and vehicle kits
- ✔ a kit for your pet, if you have one

Step 5: Prepare the House (page 35)

Annual Maintenance

RESPOND

Step 6: Respond to a Disaster (page 51)

RETURN Home

Step 7: Return Home Safely (page 67)

Let's get started!

Preface

This practical guide provides an all-hazards approach to help you prepare for any type of disaster. It consolidates research, experience, and practice into one comprehensive and straightforward plan to give you the tools you need to prepare for emergencies. Although the idea of getting yourself prepared for any emergency may seem daunting at first, the process is actually quite simple—and I'm here to escort you through that process step by step.

Why is this so important? The fact is, in this day and age, emergency preparedness is no longer just an option: it's a necessity. We must be prepared to overcome the challenges created by climate change and extreme weather events. Disasters are increasing in frequency and magnitude, with significant short- and long-term consequences for our health, our society, our economy, and the environment. On every continent, we have seen an array of increasingly frequent disasters, such as wildfires, earthquakes, hurricanes, tsunamis, heat waves, drought, flooding, disease outbreaks, and rising sea levels. An emergency can even be as simple as the power grid going down for a few days due to a storm and you needing some candles—or food and water!

In an emergency, essential services such as gas stations, roads, pharmacies, telephone and electricity service, water supplies, and grocery and hardware stores may be unavailable for days, or even weeks. While some events are relatively minor, others can be complex, with concurrent disasters creating catastrophic damage and social disruption. Disasters like this not only affect our infrastructure but can also increase poverty and lead to food insecurity and homelessness—the list of potential consequences is staggering. The loss of lives and livelihoods due to disasters across the world is devastating and sobering, and it would be unwise to entertain the notion that it won't happen to us, or that if it does, it won't be so bad. Our safety and well-being are not things that can be left to chance.

Disasters have become an important part of our reality, but we are not entirely at the mercy of their effects. We can prepare for disasters, and the fruits of such planning and efforts can pay off exponentially. And there is more good news. As I'll demonstrate in this book, this preparation is not complicated. With the step-by-step instructions in this book, getting prepared for emergencies is simple and manageable, and can include the whole family!

Effective preparedness decreases the negative impacts of disasters on our families and livelihoods, protecting our children today and giving them tools to manage the potential disasters they may experience in their lifetimes. The environment and climate have undergone tremendous changes in the last fifty years. We can only imagine what the next fifty years might look like. Only we can determine the legacy we leave our future generations—a legacy of preparedness, self-sufficiency, resilience, and survival.

We cannot leave a better legacy than a family that is prepared for its future.

Throughout my career in disaster management and public health, I've witnessed the many effects of disasters, and it has become clear to me that we are universally ill-prepared for disasters. This lack of preparation impacts our ability to manage and provide for our needs and the needs of those for whom we are responsible. With just a little effort, much can be done to help ensure the safety and livelihoods of those we love. The majority of the actions necessary to ensure this safety are simple. However, practical and reliable tools to motivate and help us prepare have been limited, which has led me to develop this book.

Together, we can build a culture of preparedness and self-sufficiency.

The first step in reaching preparedness and self-sufficiency is to adjust our mindset and know that the process of preparation is doable—anyone can do it. Following my 7 Steps, you can make preparedness part of your routine and simultaneously teach your children these essential skills. This guide makes the process easy by providing one plan that helps you prepare for all types of hazards. You won't have to flip through thirty websites and fifty brochures to determine what you need for every type of disaster or extreme weather. It is all here, made simple with easy-to-follow checklists.

The information in this book combines the latest research, science-based evidence, and practical lessons learned from past disasters, as well as my thirty-plus years of experience in disaster management, public health, and survival techniques. I've done the work and the research, and I've brought it all together so that you have everything you need in one resource to get prepared. Once you have completed the 7 Steps, you'll be ready for any disaster!

Step 1: Safeguard Important Documents

Step 2: Create a Communication and Reunion Plan

Step 3: Create a Shelter and Evacuation Plan

Step 4: Assemble Emergency Supplies

Step 5: Prepare the House

Step 6: Respond to a Disaster

Step 7: Return Home Safely

The book is laid out in several parts. In Part I: Get Ready, I outline the first five steps of the preparation process. Each step gives key information at the beginning of the chapter, followed by additional material for interested readers. In Part II: Respond, I show you how best to respond should a disaster strike. Part III: Return Home explores what happens once the disaster has passed. Finally, Additional Information acquaints you with the disaster management process, describes common hazards, and provides advice on everything from practicing proper sanitation to disinfecting water, accommodating special needs, and choosing insurance.

As you develop your emergency plan and go through the 7 Steps, make it fun and informative. Include your children, nieces or nephews, or grandchildren in the process; this could be the most important lesson you teach them. Engage them! Discuss the plan with your family and practice it so that everyone is aware of and comfortable with their role. Encourage or challenge your neighbors, relatives, friends, and co-workers to join you in your efforts. Most of all, luxuriate in the confidence and peace of mind that come *with being prepared!*

Alone we can do so little; together we can do so much. —Helen Keller

Now, enough talk! Turn the page, roll up your sleeves, and get ready to be self-sufficient!

Part I: Get Ready

Part I is the heart of this book, providing everything you need to know to *get READY*. If you're eager to dive in, start here. If you prefer to know the why and how of things before you start, or would just like a better understanding of what disasters are and what typically happens during a disaster, turn to Disaster 101 in the additional information section (starting on page 77). This section is important reading for anyone who lives in a region prone to a certain type of hazard.

Part I comprises the first five steps:

Step 1: Safeguard Important Documents (page 3)

Step 2: Create a Communication and Reunion Plan (page 5)

Step 3: Create a Shelter and Evacuation Plan (page 13)

Step 4: Assemble Emergency Supplies (page 19)

Step 5: Prepare the House (page 35)

As you read these steps, consider how to seamlessly incorporate the directions provided into your daily activities. Keep in mind that there is no need to reinvent the wheel! The forms and checklists in this book are available as free downloads from my website (details provided further in this book). You can also use the shopping list I've provided in the back of the book (see page 105) to make it simple to get the supplies you need. Schedule a block of time on the family calendar and plan your preparedness as you might plan a family vacation. Set weekly goals in your calendar or establish a friendly rivalry with neighbors or colleagues to motivate you to complete these steps. With a little imagination, you'll soon be done, and you'll be able to enjoy the ultimate reward: the wonderful peace of mind that can only come from being prepared for anything.

Step 1: Safeguard Important Documents

During a disaster, important documents may be destroyed or unavailable. It is essential to safeguard these documents and have copies of them elsewhere to have handy following a disaster.

> ### *Step 1: Gather the important documents listed below.*
> *Keep originals in a safe location that protects the sensitivity of the information, such as at home in a fireproof safety box or in a deposit box at a bank.*
> *Send copies to an out-of-town contact you trust.*

1. Gather what you need:
   - ❑ Fireproof safety box
     - These can be purchased online or from a variety of local retailers for as little as $40.
   - ❑ File folders or envelopes (optional)
     - Keeping your papers organized will reduce your anxiety when you need to find them.
     - Choosing colorful labels or markers will help keep kids engaged and interested in helping out.
   - ❑ A trusted contact's consent to hold your copies.
     - Ideal individuals include family members and close friends who do not live in your town or city.
     - Buddy up! Offer to store their important papers in exchange for storing yours.

2. Make copies of the documents listed. If you don't have access to a copier, consider going to your local library or using a print shop. Many items are easy for younger family members to help with, so enlist the troops! A downloadable list of these items is available on my website, www.authorkimfournier.com/resources/ (password is *flashlight*).

Note: In Step 2, you'll make a communication and reunion plan that will be included with these important documents.

Important Documents
- ❑ Birth certificates, proof of citizenship, adoption papers
- ❑ Marriage certificates, death certificates
- ❑ Passports
- ❑ Immunization records, medical information (e.g., prescriptions)
- ❑ Up-to-date will, power of attorney

Insurance Documents
- ❑ Life insurance
- ❑ Medical insurance
- ❑ House and vehicle insurance
- ❑ Household inventory with photos, receipts, and estimates for valuable items

Financial Documents

- ❑ Copy of credit cards (front and back)
- ❑ Bank account information (type, number), Internet passwords
- ❑ Investment, retirement, and social security information
- ❑ Tax information, self-employment documents
- ❑ Deeds, contracts, mortgages

Personal Documents

- ❑ Paper copy of address book, including email addresses
- ❑ Family history documents
- ❑ Diplomas, transcripts
- ❑ Contact information (or business cards) of preferred contractors
- ❑ Recent photos of family members
- ❑ Pet identification, pet photos
- ❑ Backup copy of important files and photos from your computer
- ❑ Step 2's communication and reunion plan

3. Place the originals in your firebox (or safety deposit box). Courier or take the copies to your trusted out-of-town contact.

4. Pat yourself on the back! You've completed the first step and are now more prepared than about 70 percent of the population!

Step 2: Create a Communication and Reunion Plan

When an emergency occurs, there may not be time to gather information and discuss a communication and reunion strategy with your family. Your family may not be at home when disaster strikes, so everyone needs to know in advance how to connect. Furthermore, local telephone service may not be available. You may not be able to communicate with your family, which can be quite stressful. For these reasons, a comprehensive communication and reunion plan is essential.

Step 2: Complete the following communication and reunion plan.

Discuss this information with your family to make sure everyone is comfortable with the plan. You may also want to test it out with a practice scenario to confirm that it works well.

The communication and reunion plan I have developed includes everything you need. Use the copy starting on page 6 or download a printable version from my website, at www.authorkimfournier.com/resources/ (password is *flashlight*). Include the following information:

1. Contact information for primary and secondary out-of-town contacts for family members to connect with in case you are separated. Choose individuals who are easy to contact and who live in another area (i.e., are less likely to be experiencing the same disaster as you).

2. Primary and alternate family reunion locations so that you can reunite with your family in case you are separated during the emergency. Your family's primary reunion place should be your home if it is safe. For alternate locations, aim for a place within the neighborhood, such as with a friend or neighbor, or at a park. Finally, choose a location outside the neighborhood. If you need to leave your reunion place, make sure you leave a note telling others where you have gone. Once your family is reunited, you can proceed to a safe shelter as noted in Step 3.

3. A list of family members, their photos, and pertinent medical information. Take new photos and confirm everyone's medical information is still accurate as part of your annual maintenance (discussed on page 48).

4. Family members' contact information. Include several ways to communicate, such as landlines, cell phones, texts, and emails, but also backup methods such as pay phones or meeting at a predetermined location. Store this contact information electronically in your cell phone and keep hard copies in several locations: with an out-of-town friend, at work, at school, at home, and in all your emergency kits.

5. Details on your pet, if you have one (or more!).

6. Medical contacts, such as your doctor and pharmacist.

7. Insurance information. Although you have the documents safe in your firebox or safety deposit box, it's helpful to have the policy details at your fingertips.

8. Vehicle information, including the make, model, and license plate number.

9. Other important numbers and contacts, such as the utility company, your lawyer, and preferred contractors.

Finally, this plan is rounded out with your community's emergency notification methods, which are available from your local emergency management office, and fire or police station.

> *Every family member should have a copy of the plan. Keep a copy in your purse, at the office, in the kids' backpacks, in your grab-and-go bag (covered in Step 4), and with your out-of-town contact.*

1. Contacts

Out-of-Town Contact (Primary)
Name:
Address:
Cell phone:
Home phone:
Work phone:
Email:
Facebook:
Twitter:

Out-of-Town Contact (Secondary)
Name:
Address:
Cell phone:
Home phone:
Work phone:
Email:
Facebook:
Twitter:

2. Reunion Plan

Family reunion place (primary shelter)
Alternate reunion place 1 (alternate shelter)
Alternate reunion place 2 (alternate shelter outside immediate neighborhood)

3. Family Member List

Name:
Date of birth:
Important medical information/instructions/allergies/special needs:
Name:
Date of birth:
Important medical information/instructions/allergies/special needs:
Name:
Date of birth:
Important medical information/instructions/allergies/special needs:
Name:
Date of birth:
Important medical information/instructions/allergies/special needs:

4. Family Contact Information

Work information for:
Workplace:
Address:
Phone:
Facebook:
Website:
Twitter:
Evacuation location:
Work information for:
Workplace:
Address:
Phone:
Facebook:
Website:
Twitter:
Evacuation location:

School information for:
School name:
Address:
School policy is to ❑ hold or ❑ release child in an emergency
Phone:
Facebook:
Website:
Twitter:
Evacuation location:
School information for:
School name:
Address:
School policy is to ❑ hold or ❑ release child in an emergency
Phone:
Facebook:
Website:
Twitter:
Evacuation location:
Day care information for:
Day care name:
Address:
Day care policy is to ❑ hold or ❑ release child in an emergency
Phone: Facebook:
Website:
Twitter:
Evacuation location:
Day care information for:
Day care name:
Address:
Day care policy is to ❑ hold or ❑ release child in an emergency
Phone: Facebook:
Website:
Twitter:
Evacuation location:

5. Pet Information

Name:	Photo:
Sex:	Color:
Breed:	
ID number:	
Important medical information/instructions:	
Name:	Photo:
Sex:	Color:
Breed:	
ID number:	
Important medical information/instructions:	
Veterinarian/kennel:	
Address:	
Phone:	
Microchip registry:	
Alternate boarding kennels:	
Pet sitter:	
Humane society:	
Poison center:	

6. Medical Contacts

Doctor:	Location/address:
Phone:	
Doctor:	Location/address:
Phone:	
Dentist:	
Phone:	
Pharmacist:	Location/address:
Phone:	

7. Insurance Information

Medical insurance company:
Name of policy holder:
Medical insurance number/policy number:
Phone number:
House/rental insurance company:
Policy number:
Phone number:

8. Vehicle Information

Vehicle make, year, model:
Vehicle insurance company and phone number:
Vehicle identification number (VIN):
Policy number:
License number:

9. Other Important Numbers

Poison control:
Police:
Fire department:
Power company:
Gas company:
Electricity provider:
Electricity repairs:
Landlord:
Phone service:
Lawyer:
Company to make structural repairs following a disaster:
Electrician:
Carpenter:
Plumber:
Other:
Report downed power lines to the power company, the fire department, and the police.

10. Local Emergency Notification Methods

Places where updated information about the situation will be posted to keep you notified. Your local emergency department will provide this information.

1. Local radio stations:
2. Websites:
3. Facebook:
4. Twitter:
5. Other:

GET READY
STEP 2

Step 3: Create a Shelter and Evacuation Plan

Disasters may cause substantial damage to your home and neighborhood, and you may need to evacuate temporarily to an alternate location. On the other hand, if your home has sustained only minor damage and is safe to live in, you may be able to shelter in your home as long as you have adequate supplies in place. Planning for emergency sheltering and evacuation is an important part of preparing for disasters.

Every disaster will affect the safety of your family, home, and community differently. Each event will dictate how, where, and when you shelter, as well as how much time you have to prepare. For example, a large earthquake may cause major damage to homes, buildings, bridges, roads, and utilities, and you will likely have little notice. A flood is generally more predictable and damage is less severe and more localized. A power outage may affect a large area but does not generally require evacuation as long as you are prepared to live without electricity. Regardless of the disaster, there are basically two options for safe shelter: shelter in place or evacuate.

Shelter in Place

Depending on the type of emergency and the time of day, you may be able to stay in your home. There may also be situations where you will have to stay at work and your children will have to stay at school for a while. In either case, it is important to think through the various scenarios that may arise, discuss them with your family, and prepare accordingly.

Sheltering in place takes on three forms: in your home, away from home, and immediate and urgent shelter in place.

1. **Sheltering in your home** may be feasible if your home has sustained only minor damage.

2. **Sheltering away from your home** might be required if your home is not safe to live in.

3. **Immediate and urgent sheltering in place** may be required if there is an urgent danger in the area, such as a release of an airborne hazardous chemical that prevents you from going outside or moving around the community. You and your family members would have to stay wherever they are during the event and take immediate shelter in place.

If your home and surrounding area are safe, sheltering in your house is generally the best option. You will have your usual comforts, experience less stress, and recover faster, and the overall situation will be more manageable than if you were forced to evacuate to a strange place, with minimal supplies.

Wherever you shelter—at home, work, or school—consider which is the safest room and what you will need to make the area more comfortable. This will vary depending on the hazard (flood, earthquake, etc.). Discuss it with your family, friends, and colleagues, and include your decisions in your shelter and evacuation plan.

Evacuate

Evacuations are more common than most people realize. Hundreds of times each year, events force thousands of people to leave their homes. An evacuation may be required if your home is not safe. It is best to understand the process ahead of time and find out if your community has plans to shelter evacuees. Develop an evacuation plan with your family that includes an escape plan out of the house, an evacuation route out of the neighborhood, and a list of supplies that you will need to bring. Use my sample form on page 15 or download a printable version from www.authorkimfournier.com/resources/ (password *flashlight*).

Even if your home is not damaged, authorities may still alert you to evacuate. Always follow instructions from these authorities, who may know of hazards that you are not aware of, to protect your family from further harm. Alerts may come in various ways: via radio, television, phone, text, Twitter, email, the Internet, sirens, fog horns, etc. Know how your community intends to alert the public in case of an emergency.

If you feel your home is unsafe to live in, or the surrounding area and buildings pose a hazard to your house (hazardous material, risk of fire, debris, unstable ground, unsafe neighboring building), you may need to evacuate outside the hazard zone. The most favorable option is to stay with a friend, neighbor, or relative. However, do not assume they will have supplies for you—even if you are staying with friends or family, don't forget to bring your grab-and-go bag (see Step 4).

If you cannot stay with a friend or relative, you may need to stay in commercial lodgings or in a mass care facility (also known as reception, group lodging, or evacuation centers). In any of these cases, moving to a different location comes with challenges and can create anxiety, especially if you are not prepared for the experience. Discuss the options with your family and prepare accordingly to minimize the stress associated with a necessary evacuation.

Part II: Respond has more details on how to shelter in place and evacuate safely, as well as tips for living in mass care facilities.

Step 3: Complete the following shelter and evacuation plan.

Discuss the plan with your family and practice it to make sure everyone is comfortable with the various escape and evacuations routes. Remove any obstacles or hazards along the routes that may restrict safe evacuation.

Post the shelter and evacuation plan in a memorable part of the house so it is accessible in an emergency and everyone can review it from time to time, such as on the inside of a cabinet door. Make a copy to put in your grab-and-go bag.

Shelter

Home Address (Primary shelter):

If your home and surrounding area are safe, sheltering in your house is generally the best option.

Alternate shelters 1 and 2 (address and name):

If your home is unsafe to live in, plan to stay with friends, family, or a neighbor.

Available shelters in your community (address and name of building or location):

Contact your local emergency manager for information on whether mass care shelters might be available and where they would be located.

Evacuation

Items to bring during an evacuation:
- ☐ Grab-and-go bag
- ☐ Water, warm clothes, waterproof coat, sturdy shoes
- ☐ Extra food, water, blankets, bedding (if you have time and space)

Escape route out of the house:

Use a separate sheet of paper or print my downloadable version. See the example and instructions below.

Evacuation route out of the city:

Use a separate sheet of paper or print my downloadable version. See the example and instructions below.

Escape Route out of the House

Make a diagram of your home's floor plan and emergency exits out of the house and from each room, as shown in the example. The diagram should include where your emergency supplies are located.

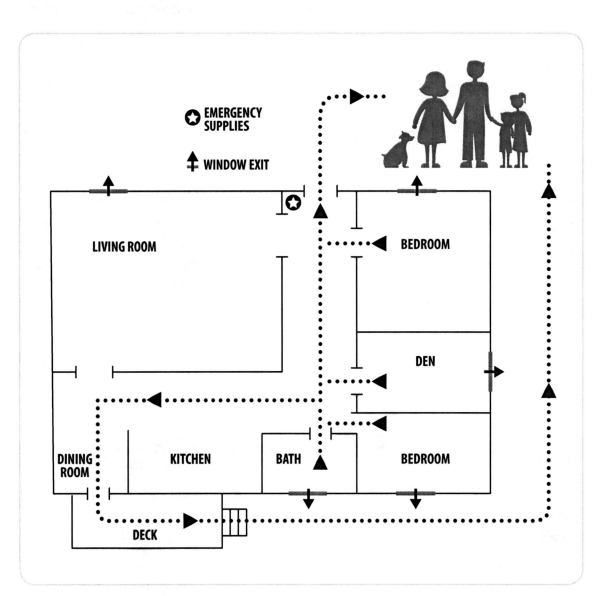

Sample diagram of an escape route out of a house.

Evacuation Route out of the City

Next, make a diagram of your neighborhood and the evacuation route out of the city. You could also print off a map from an online mapping program or your city's municipal website. You may want to contact your local emergency planners for route suggestions. Keep in mind that this plan may change based on the hazards along the route; nonetheless, it is a good idea to reflect on the possible ways out of town. Post this with the shelter plan.

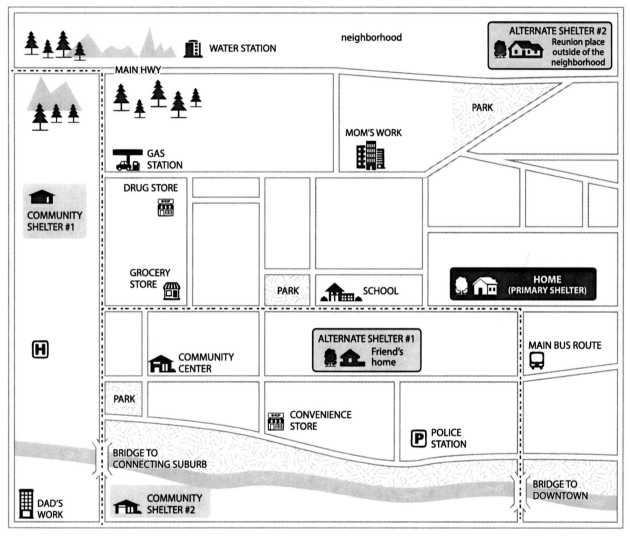

- - - - - - - - - - - - PRIMARY EVACUATION ROUTES

Sample diagram of an evacuation route out of a city.

Step 4: Assemble Emergency Supplies

Disasters can happen very quickly. When disaster hits, you might be at work, at home, at school, or somewhere in between, without access to basic necessities or services. To ensure your family's survival and comfort, it is essential to prepare the right disaster supplies for each location. As Miguel de Cervantes noted, "To be prepared is half the victory."

> ### *Step 4: Prepare your emergency supplies.*
> #### *Follow the instructions in this section.*

Emergency Supplies

Following a disaster, access to grocery stores, pharmacies, health services, hardware stores, and banks may be limited for several weeks, or longer. You may also have to live without essential services, such as electricity, potable water, telephone or cellular service, and sewage systems. Furthermore, if your home is damaged, you may need to do basic repairs, or, in a worst-case scenario, build or live in temporary shelters. Having suitable supplies for every scenario is essential to ensure your family's well-being.

You cannot know where you will be when disaster strikes, and you will therefore need emergency supplies in multiple locations. I recommend portable kits and a stockpile of supplies:

1. A kit that you can quickly take on your way out of the house, often called a grab-and-go bag.

2. A stockpile of supplies in your home, and perhaps some in your yard.

3. Kits at work and at school.

4. A kit in your car.

5. A kit for your pet, if you have one.

Assembling these supplies is one of the most important steps in emergency preparedness. Take the time to gather the appropriate items for your situation and your family's needs.

If you have children, engage their help. Kids can help to gather and pack many items. Including your kids in your preparation is a great way to alleviate fears or worries they may have and empower them to action. Their input and regular participation will also engage discussion about preparedness while encouraging them to develop a sense of survival savviness and strategies they can incorporate into their own lives as they grow.

> *"Prepare the child for the path, not the path for the child."*
> *—Native American Proverb*

If you can, prepare all the kits at the same time. This will save you time and effort. If you cannot make all the kits at once, start by assembling a kit for the location where you spend most of your time. For example, if you use your vehicle to travel to work, you could start with the kit for the vehicle. This way, you have it handy while at home and at work, and while dropping the kids off at school or extra-curricular activities. Once this first kit is complete, gather supplies for the other kits.

You may want to put the kits together over a weekend or spread the task out over several days or weeks. People often mention that they've wanted to prepare emergency kits for many years but have never found the time. While it can be hard to *find* time, it's just a case of rearranging priorities to *make* time. Once you have prioritized preparedness and have dedicated time to arranging your emergency supplies, commit to maintaining your kit and emergency plan on a regular basis.

You can either make the kits from scratch or buy pre-made kits and add extra supplies. You likely have many of the items you need already around the house. Store-bought kits are typically not adequately stocked and always require additional items to ensure enough provisions to meet your family's needs for food, personal hygiene, water, health, and safety. If you can, aim for three weeks' worth of supplies or more, and follow my plan to restock items as required.

Remember that making emergency preparation a priority is essential to protect your family and ensure their survival in case of disaster. This is an important responsibility; preparedness is one of the most important gifts you can give yourself and your family, so enjoy the process. Once you've gone through the process yourself, encourage your siblings, friends, colleagues, and neighbors to join you in this important task and prepare their families for inevitable disasters.

Below is a list of recommended supplies for your grab-and-go bag; home supplies; and work, vehicle, and pet kits. Remember to tailor the items and quantities to your family's situation, needs, environment, and local hazards.

Grab-and-Go Bag

You will need a bag (or bags) with emergency supplies that you or other family members can quickly grab if you need to evacuate. The grab-and-go bags should have enough supplies for the whole family for at least three to seven days.

Follow the tips below to assemble your kit. Always adjust items based on your preferences and your family's needs. If you cannot purchase all the supplies at once, prioritize items based on your needs for water, food, shelter, protection of life, and safeguarding from illness or injury.

Choosing the Grab-and-Go Bag

- The bag should be sturdy, easy to carry, and waterproof (or at least water repellent).

- If you do not have a water repellent bag, you can line the inside of your pack with a large garbage bag. The lining will help protect your supplies while also serving as an extra multi-purpose bag you can use later.

- A bag with many compartments and outside pouches is most practical. The exterior pouches can store items you will need accessible for immediate use.

- A backpack is handy as it leaves your hands free for other tasks. You may need your hands to remove debris, carry a child or pet, bring additional supplies, or help others evacuate. An alternative option for a grab-and-go bag is a gym bag with a shoulder strap or a suitcase on wheels. However, keep in mind that it may be challenging to wheel a bag if there is debris on the streets.

- A small wagon can also be used to carry your gear, especially since the solid tires cannot be deflated by sharp debris.

- You may opt to have one large pack for the whole family or several smaller ones for each person; choose the option that works best for you. Remember that not everyone will be home during the disaster and may need to meet you at a different location later. Therefore, all the necessary supplies for the whole family should be portable.

- Keep an inventory of supplies inside the pack to make annual maintenance quick and easy. Include the expiration dates of food and medication.
- Place two waterproof identification tags on the outside of the bag, such as those used for travel luggage.
- One tag should include the name of the person who will use the pack. If you have one large bag for your family, include everyone's name.
- The second tag should include a checklist of the following reminders to be consulted before you leave the house. This is important: In the rush of evacuating, you may not remember to bring critical items that will serve you later.

  - ✔ Does everyone have warm and waterproof clothes, and good footwear?
  - ✔ Do you have the car and house keys, your cell phone and charger, everyone's identification, and your wallet or purse?
  - ✔ Do you have the time and space to bring more water, extra warm clothes, rain clothes, and non-perishable food?
  - ✔ Is the first aid kit securely attached to the outside of the bag with a carabiner, rope, or triangle bandage?

Stocking and Storing the Grab-and-Go Bag

- Keep the grab-and-go bag near an exit that you will most likely evacuate from, such as in a closet near the front door. Avoid planning to leave from the garage in case there is a power failure and the garage door won't open.
- When evacuating, you want to bring as many supplies as possible. You may even consider bringing a portion of your home emergency supplies with you when you leave home. For this reason, you may consider storing the grab-and-go bag close to your home supplies.
- You may also want to keep an empty bag near your grab-and-go bag in case you choose to bring extra food, water, or clothing from the house.

The items below can be stored inside the grab-and-go bag. Adjust quantities depending on whether you have one large pack for the whole family or several smaller ones for each person. Specialty items are available at most safety, camping, or outfitter stores, or search for them online.

Items for the outside pouches

- ❑ Cash in small bills and coins
- ❑ Car and house keys (yours and a friend's)
- ❑ Spare cell phone charger
- ❑ Non-latex disposable gloves

- ❑ Pen, pencil, pad of paper
- ❑ Poncho or raincoat (a poncho is multi-purpose and can be used as an overhead cover or small tent)
- ❑ Heavy duty work gloves
- ❑ Alcohol-based hand sanitizer, tissues
- ❑ Sunglasses if you are sensitive to sunlight
- ❑ An old pair of prescription glasses
- ❑ Flashlight or headlamp
- ❑ Copies of your up-to-date communication, evacuation, and shelter plans

Note: Remember to keep all electronics and items that may get damaged by water in resealable waterproof bags.

Items for personal hygiene

- ❑ Medication you cannot live without for a few days
- ❑ Multivitamin, over-the-counter painkiller
- ❑ Toothbrush, toothpaste, dental floss
- ❑ Multi-purpose soap, shampoo, towel, deodorant, lip balm
- ❑ Feminine hygiene supplies
- ❑ Face cream, sunblock, bug spray
- ❑ Prescription glasses
- ❑ Flip flops for public showers
- ❑ Toilet paper
- ❑ Powdered laundry soap (avoid liquid soap, which may freeze if the weather is cold)

Items for protection and warmth

- ❑ Hat, coat, sturdy shoes, or winter or rubber boots
- ❑ Gloves or mittens, warm hat, scarf
- ❑ Spare clothes (long pants, underwear, thermal undergarments, socks)
- ❑ Emergency blanket (silver, heat-reflecting)
- ❑ Sleeping bag or blanket (an airplane blanket is compact and warm and is a great option)
- ❑ Tent (or plastic sheeting)

Safety items

- ❑ N95 mask (to protect from dust, debris, or pandemics)
- ❑ Candles (for heat, light, or to start a fire)
- ❑ Extra batteries for flashlight, headlamps, and other devices. You may want to store batteries outside your devices. Standard alkaline batteries can rupture and leak chemicals, causing damage. Some devices also maintain a current, which will eventually drain the batteries. Newer lithium polymer batteries have a shelf life of approximately five years and don't lose their charge significantly, but they can drain over time.

- ❑ Waterproof whistle (the international signal for help is three short blasts)
- ❑ Portable, battery-powered, or wind-up radio with extra batteries
- ❑ Small sewing kit (such as those found in hotel rooms)

Utility items

- ❑ Rope (parachute rope is excellent; it is light, strong, and compact, and the smaller inner threads are handy)
- ❑ Zap straps (zip ties), duct tape
- ❑ Multipurpose tool
- ❑ Safety goggles
- ❑ Resealable storage bags, large garbage bags
- ❑ Waterproof matches, lighter
- ❑ Knife, hammer, multi-screwdriver, screws, nails
- ❑ Two-way radios (walkie-talkies). Depending on your situation and where your family spends the most time, you may have two sets. With the first set, put one radio in the grab-and-go bag and another in the vehicle kit. With the second set, put one radio in your oldest child's school kit and the other in your work kit.
- ❑ A copy of this book for easy reference
- ❑ Disposable camera (waterproof)
- ❑ Compact survival book

Items to support mental wellness

- ❑ Pictures or comfort items
- ❑ Pocket-sized game or deck of cards
- ❑ Toys or stuffed animals for the kids, coloring books, pens or pencils and paper
- ❑ Music, books, games, teddy bear
- ❑ Earplugs (you may be sleeping in a crowded area)

Infant needs

- ❑ Nutrition for at least three to seven days (check with your pediatrician to ensure proper nutrition in an emergency)
- ❑ Extra clothing
- ❑ Diapers and wipes for three to seven days
- ❑ Blanket
- ❑ Favorite stuffy
- ❑ Medication
- ❑ Copy of health and immunization records

First aid kit

A first aid kit is essential and should include items for wound care, sprains, and fractures. Store the first aid kit for easy and quick access, such as in an outside pouch. If you do not have room in your bag, attach it to the outside using a carabiner, rope, or triangle bandage. The first aid kit should be attached so it doesn't fall from the bag but is handy to detach if you need to treat an injury. Here are some suggested items to put in your first aid kit:

- ❑ Non-latex gloves (in addition to the ones in your outside pouches)
- ❑ Bandages, gauze, elastic bandage (tensor bandage)
- ❑ Medical tape
- ❑ Face masks
- ❑ Slings and splints
- ❑ Cotton tipped swabs, disinfectant
- ❑ Sanitary napkins (these are multi-purpose, very absorbent, and excellent for large wounds)
- ❑ Multi-purpose scissors, tweezers
- ❑ CPR mask
- ❑ Emergency blanket (in addition to the one in your outside pouch)
- ❑ Medications (anti-nausea, anti-diarrhea, antihistamine, analgesics or painkillers, antibiotic ointment)
- ❑ Pocket-sized first aid manual

Note: Everyone should be comfortable with first aid procedures. If your adult family members and older children are not, enroll them in a first aid course. This knowledge is the perfect gift for a special occasion.

Food and water

- ❑ At least three to seven days' supply of food
- ❑ Basic cooking supplies (manual can opener, ladle, cutlery, collapsible bowls and cups, plates, utensils)
- ❑ Alternative cooking stove
- ❑ At least three to seven days' supply of water for drinking, cooking, and hygiene (can be carried separately from the grab-and-go bag)
- ❑ A chlorine tablet product to disinfect water in case boiling is not available to you

Note: See the following sections for suggestions on choosing food and cooking items, determining how much water you need, and preparing and storing your water. In the additional information section at the back of the book, Collecting and Disinfecting Water (beginning on page 89) includes information on disinfecting water taken from alternate sources such as lakes, rivers, rainwater, and ponds.

Disaster Supplies for the House

Not all disasters will require you to evacuate. You may need to shelter in your home with no electricity, potable water, or a working water system for some time. Your house may have sustained damage that requires repair, such as broken windows, holes in the walls, leaks in the roof, and build-up of debris. The disaster supplies you keep in your home should include items to support you and your family to live in these situations for at least three weeks. Have enough food and water for an extended period of time, as well as alternate lighting, shelter, and heating, and basic home repair items.

In addition to the items included in the grab-and-go bag, the following are a few suggestions to add to your home disaster supplies or have ready around your home.

Type and Location of Home Kit

- Portable supplies should be kept near an exit, such as in a closet or cabinet.
- Use a waterproof or airtight container. Another option is to use one or two easy-to-carry containers (clean trash can on wheels, duffle bag, suitcase, or backpack) and place the supplies in the containers in plastic bags.
- Store supplies in a cool, dry place. Perishable foods will last longer if they are stored this way.
- Avoid storing supplies on concrete floors, which may cause humidity to build up and increase corrosion of cans.
- Store supplies away from fuel or toxic chemicals.
- Keep supplies free from debris and do not let them get covered by other items.
- Keep an inventory with the supplies, including expiry dates and the date that you last checked everything. This will help for annual inventory checks.
- If you have a camper or trailer, you may already have a good supply of emergency items.
- Some people find it convenient to keep supplies in one place so that they are easier to locate or gather in an emergency.

Items for the Home Kit

- ❑ Face masks
- ❑ Alcohol-based hand sanitizer
- ❑ Several candles
- ❑ Larger flashlights (consider non-battery-powered options like solar or wind-up)
- ❑ A first aid kit with more quantities of items than the one in your grab-and-go bag (can be bought ready to go)
- ❑ Additional clothing
- ❑ Additional food, water, and cooking supplies (cups, utensils, dishes)
- ❑ Generator and additional fuel
- ❑ Garden hose for siphoning and fire fighting
- ❑ Emergency sanitation and toilet (see Hygiene and Sanitation starting on page 83 for details).
- ❑ Household chlorine. Always have a supply around the house to disinfect surfaces and equipment in the event of a pandemic. As well, chlorine can be used to disinfect water.

Items for Home Repairs

- ❑ Brooms and shovels for removing debris
- ❑ Tarps
- ❑ Hacksaw, axe, crowbar; wrench to shut off gas and water valves
- ❑ Sheets of plywood for minor or urgent repairs to damage to the house
- ❑ Vapor barrier for various repairs (cover broken windows, seal off holes in the exterior walls, seal an interior room for warmth)
- ❑ Hammer, screwdriver, various nails and screws

Items for Sheltering

- ❑ Sleeping bag. If you have a sleeping bag in your grab-and-go bag, you can take it from there. However, if you opted to have portable blankets in your grab-and-go bag, you may want full-sized sleeping bags with your home emergency supplies.
- ❑ Cots
- ❑ Air mattresses
- ❑ Tent or tarp. In cold weather, you can set up a tent indoors or close off one room to maintain body heat in one area.
- ❑ Supply of firewood (if you have a wood stove or fireplace)

Food and Water

Include enough food and water in your home disaster supplies for at least seven days—aim for a three-week supply. I provide detailed instructions in the following sections.

Food

Food is vital to our well-being. Food also lifts our morale and gives us a feeling of security during times of stress. Although it is difficult to predict how much food you will need for emergencies and how difficult it will be to replenish your supply, it is best to stock more food than you expect to need.

Grocery stores typically maintain only a few days' worth of supplies to avoid spoilage and reduce inventory costs. For this reason, grocery stores will quickly run out of stock following a disaster and have challenges restocking because of a compromised resupply chain (damaged roads, broken delivery trucks, etc.). Suppliers will face similar challenges. For these reasons, it is paramount to keep extra food in the house and in your emergency kits. Keep three to seven days' worth of food in your grab-and-go bags and at least seven days' worth at home—preferably three weeks' worth or more.

To determine the right quantity of food, track your family's consumption during the week and weekend. Then, slowly stockpile your food supplies. Purchase a few extra items every week when things come on sale. Try to keep extra food in the cupboards and replace it once consumed. Know that if physical activity is reduced, healthy people can generally survive a few days without food, and they can get by on half of their normal food intake for an extended period.

When stocking your emergency food supply, aim for foods that…

- Your family enjoys
- Require a minimum of cooking, preparation, fuel, refrigeration, and water
- Are lightweight and compact
- Do not increase thirst—low-salt foods are best
- Accommodate your dietary needs
- Have a long shelf life
- Are as high in calories and nutrition as possible

Note: You may want to explore some sports and outfitter stores for ready-to-eat food options. They also have foods with a five-year shelf life, as well as self-heating pouches to warm up the food. Children often love the idea of foods such as freeze-dried ice cream and can help by choosing canned goods they like, so include them as you shop.

Suggestions for foods

- Canned beans, lentils, legumes
- Marinated or canned vegetables
- Canned fruit, soup, pasta, chili
- Quick-cooking dried noodles, rice, oatmeal
- Protein bars, dried fruit, nuts, crackers
- Tea, coffee, powdered milk, spices, peanut butter
- Dehydrated meals or ready-to-eat foods, such as those purchased at camping stores
- Comfort foods—candies, gum

Cooking

- Include an alternative cooking source in your kits. Examples include candle warmers, fondue pots, camping stoves, and charcoal grills. If you aren't forced to evacuate, you can use your fireplace or barbecue.
- Keep your barbecue tank full and have two full propane tanks on hand. During a disaster, the local propane supply will quickly run out.
- Always operate stoves safely and never use charcoal or barbecue stoves or heaters indoors.

Include the following cooking items in your kits:

- ❑ Different sizes of candles
- ❑ Manual can opener, utility knife
- ❑ Ladle, spoons, forks, knives, bowls, plates, cups (collapsible cups and bowls are handy)
- ❑ Waterproof matches
- ❑ Oven mitt (or you can use a glove)
- ❑ Collapsible dishwashing container (can also be used for personal hygiene)
- ❑ Dish detergent, pot scrubber, bleach

Food storage

- Light, heat, humidity, oxygen, and pests can make food deteriorate more quickly. Store food in airtight bags or closed containers and place in a cool, dark place, away from potential access from pests.

- Discard foods that have come into contact with floodwater or that have an unusual odor, color, or texture.

- Do not eat food from cans that have been damaged, punctured, swollen, or corroded, or that look or smell abnormal.

- Discard unused instant cereal, peanut butter, jelly, and canned foods after one year, regardless of the expiry or best-before date.

- Discard powdered milk, dried fruit, and crackers after six months.

- Consider dehydrating or canning food yourself.

- Rotate the food that you store in your cupboard.

Nutrition tips

- Eat at least one meal a day.

- Remember that the first few days of rationing food may be the most challenging, but your body and mind will adjust quickly. Try not to focus on the lack of food; rather, appreciate every bite.

- Drink enough liquids to allow your body to function properly. A good idea is to monitor your urine. If you find you are barely urinating, or if the urine is dark yellow and concentrated, you may need to increase your water consumption.

- Consider taking a daily multivitamin to supplement your nutritional intake.

Water

Water is essential for life and is one of the most important items in your emergency supplies. We cannot live without water! You will need a supply of water for drinking, personal hygiene, and cooking in your grab-and-go bag; vehicle, work, and school kits; and home emergency supplies.

International guidelines for water requirements in survival situations recommend two to three quarts (two to three liters) of drinking water per person per day, one half to one and a half gallons (two to six liters) for personal hygiene, and one to one and a half gallons (three to six liters) for cooking. However, individual needs depend on the person, age, activity, diet, health, and climate.

Most people do not normally drink that much and may only consume one or two quarts (one to two liters) per day. Likewise, depending on the foods you include in your emergency supplies, you may need only one to two quarts (one to two liters) per person per day for cooking and the same for personal hygiene: a total of one to one and a half gallons (three to six liters) per person per day, including what you need for drinking. The best strategy to know exactly how much you will need is to monitor your family's water consumption over a weekend and a few days during normal workdays, and adjust your supply accordingly.

Your grab-and-go bag should have three to seven days' worth of water for drinking, cooking, and hygiene. Your bag will be heavy to carry with adequate amounts of water, so you may need to be creative, such as putting the water in a wagon, a rolling cart, or a suitcase with wheels. Water containers are available in various sizes at camping, hardware, or outfitter stores.

Your home supplies should include at least a seven-day supply of water—aim for at least three weeks' worth. Remember, your hot water heater can supply over forty gallons (150 liters) of water as long as it has been secured to withstand the hazards in your area. See page 87 for steps on how to collect and disinfect the water from your tank. Also, large drums or tanks are commercially available at many safety or survival stores.

Aim to stock enough supply that you will not need to collect water from alternate sources, such as a lake, river, pond, or rainwater. However, if water supply is cut for an extended period, you should have a plan and know how to collect and disinfect water from alternate sources (instructions on how to do so start on page 89).

Don't forget about your pets! Include up to one quart (one liter) of water per day for your pet. You can monitor your pet's consumption to have an exact measurement. Keep in mind that during a disaster, your pet may be particularly stressed and pant more, which will require more hydration.

I cannot emphasize enough the need for adequate quantities of water. It is critical to acknowledge this need and prepare accordingly. It is quite easy to set up and maintain a system to ensure you have enough water for your family. This is one of the most important activities you can do to secure your family's health and well-being in a disaster response.

Preparing and storing water

Properly collecting and storing water is paramount to protect your supply and ensure good water quality when you need it. Consider the best method for you and the optimal place in the home to store the water.

Other than alternate water sources, there are two main ways you can obtain water for your emergency kits. You can either buy commercially bottled water or fill containers from your local water supply.

If you purchase commercially bottled water for your kits, the containers must be kept closed and sealed, and, optimally, rotated monthly. Set a reminder on your calendar or smartphone. Commercially bottled water is packaged in plastic that is not meant for long-term storage and can deteriorate over time. Furthermore, the amount of disinfection in the bottles is typically insufficient to sustain potability for longer periods. Unless you are diligent in rotating the bottles monthly, you may prefer to prepare your own water using food-grade containers filled with your local water supply.

If you prepare your own water, use high quality food-grade bottles. Avoid using glass bottles, as they are heavy and can break easily.

- Wash the bottles with soap and water and rinse well before filling with potable water. Then add a disinfectant of 1/8 teaspoon of household chlorine bleach per half gallon (two liters) of water and swish around to ensure the sanitizer contacts all the surfaces of the bottle. Allow the sanitizer to sit for thirty minutes to completely destroy pathogens. Then empty the water.

- Rinse thoroughly and fill with potable water. Do not fill to the top, so as to allow room for the water to expand or freeze. Once filled, you can add a few drops of unscented bleach to extend disinfection and ensure potability for up to three months. After three months, replace the water.

- If you do not add extra chlorine, it is important to change the water monthly; otherwise the quality decreases as the chlorine dissipates, promoting bacterial growth. Shake the bottles monthly to ensure the entire supply is disinfected.

Additional water tips

- If you have space in your freezer, it is a good idea to freeze water in large resealable baggies; these can be used to preserve perishable food if you lose electricity.

- You may also add a few emergency drinking water packets to your kits as a backup. These packets are available at most survival or safety stores. They typically contain only about two cups of water and are not cheap, but they have shelf life of five years as long as the package is not damaged and is not stored in extreme temperatures. These packs may break if they freeze.

Alternate sources of water for drinking and washing

There may be alternate water sources in and around your community. Look around your area to see if there is a source, such as a lake, river, or pond within walking or biking distance. Have a plan and supplies ready to collect, transport, and disinfect the water. Test your plan to make sure it is feasible and you are comfortable with the process.

- **Safe water sources for drinking:** Melted ice cubes, liquids from canned goods (fruit and vegetables).

- **Water sources that are safe once filtered and/or disinfected:** Toilet tank (not toilet bowl), hot water heater, rainwater, streams, rivers, ponds, lakes, and natural springs.

- If you are in doubt as to whether the water is safe to drink, disinfect it.

- **Unsafe water sources:** Radiators, waterbeds, swimming pools and spas, water from toilet bowl. Water from pools, spas, and waterbeds is unsafe to drink as it has a high concentration of salts and chlorine.

- **Water sources for washing:** Hot tub, pool.

Note: If you hear of broken pipes or sewage lines, you will want to protect the water sources already in your home until you are assured the water is safe to drink. Do so by shutting off the main valve for the water entering your house; see Step 5 for detailed instructions.

Collecting rainwater

Collecting rainwater is an excellent way to conserve resources and augment drinking water supplies at any time—and especially when disaster strikes. Rainwater collection is not allowed in all areas, so check your local policy.

Rainwater is not clean and can carry bacteria, parasites, viruses, and chemicals, so you must collect the water in such a way as to avoid extra contamination. Collect rainwater in a sealed, clean barrel, which can be placed in an open field or under your eaves. Install it in such a way that children cannot accidentally get in. Cover the opening with a fine wire mesh to prevent leaves, debris, bugs, bird droppings, and rodents from entering the barrel. Filter and disinfect the water before consumption (see Collecting and Disinfecting Water starting on page 89).

Work, School, and Vehicle Kits

- These kits should each be in one container or a bag, placed in a convenient location such as under your desk or in a locker, ready to go.

- The inventory should be similar to the grab-and-go bag (adjust items and quantities as required).

- If you typically wear dress shoes or high heels at work, you may want to include a sturdy pair of shoes in your work kit in case you have to evacuate or walk for a long distance.

- You may be stranded somewhere and require supplies to survive on your own. Have a kit handy in all of your vehicles, including RVs and boats.

- Keep your vehicle in good repair and always keep your gas tank at least half full.

- The vehicle kit inventory should be similar to the grab-and-go bag (adjust items and quantities as required). Other than your basic vehicle supplies (spare tire, booster cables, etc.), add the following items:
  - ❑ fire extinguisher
  - ❑ flares
  - ❑ seasonal supplies such as an ice scraper and a small shovel

Preparing Your Pet for a Disaster

If you have a pet, it is doubtless an integral part of your family and everyday life, so it's important to remember it when planning for a disaster. Thinking ahead will help improve your pet's safety and chances of survival. Whether you shelter at home, with a friend, or in a reception center, there are several things to consider to ensure your pet makes it safely through a disaster or emergency.

Depending on the disaster, a few scenarios may occur:

- If your home is safe to live in, you can keep your pet at home with you.

- If your home is unsafe and you stay with a friend or family member, always bring your pet with you.

- If you evacuate to a reception shelter that does not take animals, you will need to find an alternate shelter for your pet.

- As a last resort, if you cannot find shelter for your pet, you might have to leave it at home.

Supplies for Your Pet's Disaster Kit

Place your pet's items in a separate bag that can be attached to your main evacuation kit. You may evacuate with or without your pet and should have supplies ready for either scenario. You may also need to leave your pet with a friend or at a shelter.

Here are some suggested items to include in the kit:

- ❑ Recent photo of your pet with the family
- ❑ Leash or harness, tied on the outside of the bag for easy access
- ❑ Pet food (aim for the most concentrated and smallest bite size)
- ❑ Can opener (if you bring canned food)
- ❑ Water and a drinking or eating container
- ❑ Two weeks' supply of pet medication; include written instructions

- [ ] Towel or blanket
- [ ] Copy of your pet's recent vaccination record and spay or neuter certificate
- [ ] Pet first aid kit (hydrogen peroxide, bandages, appropriate pain medication, pet first aid booklet, small splint, etc.)
- [ ] Poop bags, paper towels, hand sanitizer
- [ ] Toy
- [ ] Litter box, litter, scoop (if you have an indoor cat)
- [ ] Cleaning supplies, preferably disposable bleach and soap wipes, to wash crate, food and water bowls, and litter box
- [ ] Emergency phone numbers. Include your phone number and address (also include a secondary contact), emergency pet clinics, animal shelters, boarding kennels, pet-friendly hotels, and your veterinarian's number and address.
- [ ] Pet carrier. Depending on where you shelter your pet, you may consider bringing a pet carrier. If you don't use it regularly, train your pet to be comfortable with this form of transportation.
- [ ] Pet booties. In the event your pet has to walk through sharp debris, booties can spare your pet discomfort or injury. Practice putting the booties on your pet from time to time so your pet is used to wearing them.

Evacuating with Your Pet

- Bring your pet disaster kit.
- If there is sharp debris on the roads, consider equipping your pet with protective booties.
- If you are evacuating to a place that does not take pets, contact the pre-arranged backup location (boarding shelter) to ensure it is still accepting animals.

Leaving Your Pet Home Alone

- Tag your pet with proper identification.
- Leave current vaccination records and any other pertinent information in a visible place in the house, such as on the kitchen counter. This may come in handy if a friend checks in on your pet.
- Leave enough dry food and water for the estimated time you will be away.
- Ensure there is adequate ventilation in the house.
- If you have time, remove any hazardous items or liquids around the house that your pet may get into.
- You may consider confining your pet to one room for safety.

General Tips

Disasters will confuse and frighten your pet; be aware of your pet's and other animals' behavior. They may be more protective, nervous, aggressive, or sensitive to noises. They may not understand what is happening, but they will definitely sense something is wrong. They may become uncomfortable and disoriented without their normal scent markers to guide them. Pets might also get frightened and wander off. At the first sign of an emergency, put a leash on your pet and bring it indoors, if it is safe. After the disaster, watch your pet closely for any unusual or aggressive behavior.

- Ensure your pet has a current identification tag on the collar, including the pet's name and your contact information.
- Have a current photo of your pet, showing color and size for easy identification; include your family in the photo to prove ownership.
- Consider having a pre-made lost pet poster in your emergency kit.

Have a plan and supplies ready in case you must evacuate, and pre-arrange a few alternate places for your pets to stay in case you cannot take them with you. You might consider a friend's place or boarding shelters. If you must travel outside with your pet, be aware that disasters can cause many hazards for animals, such as sharp debris, fuel spills, and spilled chemicals or fertilizers that can be toxic—protective booties can help keep your pet unharmed.

Step 5: Prepare the House

Preparing your home to withstand disasters and extreme weather events is an important and worthwhile exercise. Such preparation will minimize injury, reduce stress, decrease damage to your house and belongings, reduce financial loss, and improve recovery so that you can return to your normal routine more quickly. If you live in a multi-residential complex (a townhouse, apartment, or condo), please review the special considerations I outline at the back of this section.

> ### *Step 5: Prepare your home, family, and pets for disaster.*
> *There are many action items in this step. Follow the instructions that are relevant to your living situation.*

General House Preparation Guidelines

Severe weather and disasters can create a number of hazards around the home, on the roads, and in your community. Some severe weather will displace loose objects, such as garbage cans, shingles, lawn furniture, barbecues, or tree branches. These items can move forcefully, injuring people and damaging property. Here are some tips to prepare for severe weather and other disasters.

General Logistics

- Keep your cell phone and tablet charged.
- If you have a landline, ensure you have at least one corded phone in the house. Cordless phones do not function without electricity.
- Always keep a bicycle handy in the event your vehicle is not serviceable.
- Keep a chain, lock and key, tire repair kit, and air pump on the bike.
- Keep your vehicle in good repair and your gas tank at least half full.
- Keep your vehicle in a garage or carport, if possible, to prevent damage from fast-moving objects.
- If you heat with oil, keep your tank at least half full at all times.

Around the House

Outside

- Maintain your house in good repair (roof, eaves troughs, sump pump, etc.). It's a good idea to walk around the house once a year with a focus on disaster preparation. I've provided many tips in the annual maintenance section starting on page 48. Set a date on the calendar so you don't forget!
- Secure loose articles so they don't blow around or get torn away.
- Secure toxic chemicals, preferably in a locked, waterproof container, or on a shelf a few feet above the ground in a shed (in case of flooding).
- Keep the yard tidy; store loose items in a shed. Loose objects, such as lawn furniture, can fly around and injure people.

- Trim branches and dead trees to reduce the risk of them falling on your roof, vehicle, or power lines.

- Adjust the drainage around the house to reduce the risk of flooding during snow melt or heavy rains.

- Make sure you have snow shovels handy, even if you live in a location that does not typically get snow. Stock up on ice-melt salt or sand to help reduce the slipperiness of ice.

Inside

- Have necessary devices to escape safely (safety ladders on balconies and near second-floor bedroom windows) and ensure the windows open fully.

- Have an axe in the attic in case you must evacuate from there.

- Have your emergency kit ready to go and flashlights handy. It is a good idea to keep sturdy footwear (with warm socks!) and a flashlight under every bed in case you must evacuate quickly at night.

- Have extra flashlights hanging on several door handles around the house.

- If you have a fireplace or wood stove, ensure you have enough firewood to heat and cook for several weeks.

- Always have household bleach handy to disinfect water or contaminated items, or to clean surfaces in the event of a pandemic.

- Have recreation vehicle antifreeze in case you have to prepare the house for a freeze-up.

- Be prepared to live without electricity for a while (see Step 6 for details).

Prepare Your Home for a Wildfire

Wildfires have occurred naturally for thousands of years, but their frequency and magnitude are increasing, likely due to human carelessness, population growth in closer proximity to forests, and changes in weather trends. Wildfires can greatly disrupt a community, causing extensive damage. Preparing your home, property, and business for such an event can significantly reduce the risk of damage and loss. The following are general safety tips to reduce the risk of your home catching fire and lessen the damage from a wildfire:

- Reduce the use of any materials that may fuel or ignite a fire on your property and in your home.

- Remove any trees, shrubs, and woodpiles within thirty feet (ten meters) of your house. Keep your grass mowed and watered.

- Aim for fire-resistant plants, groundcover, and noncombustible and fire-retardant or -resistant building materials.

- Keep your property and house (including roof, eaves, and rain gutters) tidy and clear of debris, dead branches, and heavy vegetation.

- Keep garbage and flammable materials away from ignitable sources such as furnaces, water heaters, and wood stoves.

- Keep a garden hose handy to put out small fires.

- Have a shovel to help prevent fire from spreading.

For more detailed information about how to prepare your home to reduce the risk of wildfires, please visit the following websites.

- In Canada: *Homeowner's Manual* from the FireSmart Canada website at http://www2.gov.bc.ca/assets/gov/farming-natural-resources-and-industry/forestry/wildfire-management/prevention/prevention-home-community/bcws_homeowner_firesmart_manual.pdf
- In the United States: http://www.ready.gov/wildfires
- Visit Smokey the Bear, now over 70 years old, at http://www.smokeybear.com/wildfires.asp

Prepare Your Home for a Flood

Floods are one of the most common and most costly disasters. Floodwaters can move fast and can be very powerful and destructive. Preparing for floods can minimize damage and personal risk. The following are suggestions to prepare for a flood.

- Keep storm drains and gutters clean.
- Ensure landscaping design and downspouts drain water away from the house.
- Make sure your insurance includes flood protection.
- If you live in a flood zone, move dangerous chemicals such as weed killers, insecticides, and corrosives to dry areas to prevent chemical contamination, fires, and explosions.
- Install a sump pump, check it monthly, and make arrangements for backup power.
- During heavy snowfall, shovel snow away from your property.
- If you live near a waterway, keep updated on the weather, as well as the risk of rising waters and floods. Be prepared to sandbag and have enough supplies to make a one-foot (thirty-centimetre) wall or dike around your property. Supplies include sand, burlap, plastic bags, and shovels.

Prepare for a freeze-up

If your water pipes are at risk of freezing, you will have to prepare the house for a freeze-up. Your pipes may freeze if the interior temperature of your house drops below freezing for an extended period of time. Here are a few potential situations where you might have to prepare the house for a freeze-up.

1. You must evacuate your house in the winter.
2. You do not have electricity or an alternate source of heat for an extended period of time.
3. You do not have water coming to your house during the cold of winter.

Follow these steps to prepare for a freeze-up:

- Drain the water pipes to prevent them from breaking. This can be done in two steps:
    - o Shut off the main water valve (where the city water comes into the house).
    - o Open all the taps in the house, including the basement, to drain the pipes; collect the water in containers for later use.
- Drain the hot water tank, furnace humidifier, washing machine, dishwasher, and hot tub.
- Add recreation vehicle antifreeze to all the sinks, bathtubs, showers, and floor drains.
- Protect the valve and pipes around the main water valve as much as possible by covering them with a blanket or insulating material.

Prepare Your Home for an Earthquake

During an earthquake, the ground will shake and may cause damage to your property. If the earthquake is large enough, the shaking may damage buildings, bridges, and tunnels; disrupt utility services (phone, water, sewer, gas); and trigger landslides, fires, and floods. However, when families and communities are prepared, the impact of an earthquake can be significantly reduced. Preparing your home and its contents for an earthquake can prevent injury and property damage, and is well worth your time and financial investment.

Homes that are built from wood seem to be the most resilient to earthquake damage. Conventionally framed houses are less likely to collapse, as long as they remain on their foundations and the roof, ceiling, and walls remain connected. However, these buildings are not entirety immune to the effects of an earthquake.

Buildings generally withstand vertical forces but are not as well protected from the horizontal motion caused by earthquakes. Homes built before the seismic upgrades to building codes may not be bolted to their foundations. Homes built after the upgrades may be bolted to their foundations but possibly improperly secured. In the case of an earthquake, this can cause structural damage and result in broken utility connections.

Understand the risk of damage to your home so that you can plan accordingly and secure your belongings to minimize injury and loss. Have a solid evacuation plan, consider additional insurance, and explore the need, cost, and benefit of retrofitting your house.

Brick and masonry

Brick, masonry, and stone facades are susceptible to earthquake damage. In the event of an earthquake, family members should keep away from such facades to avoid injury from debris. If your house has a lot of brick or stone, you may want to consult a structural engineer for advice on how to secure it.

Chimneys often collapse during earthquakes. Loose bricks become a hazard if they penetrate the roof and fall to the rooms below, or outside. You may want to inspect the chimney for loose tiles or bricks and reinforce the ceiling surrounding the chimney to provide protection from falling bricks that might break through the roof.

Windows

Broken windows are a significant hazard in an earthquake. You may consider adding safety film to the inside of any windows that are greater than two square feet (0.2 square meters). This does not prevent the window from cracking, but it does keep the glass from falling and injuring anyone. Closing the curtains and blinds in the evening can help to contain breaking glass and keep it from flying around.

Furniture and belongings

A major earthquake can cause substantial damage to your belongings. The value and location of your furniture will dictate how you secure it. There are many ways to protect each piece. Here are a few ideas.

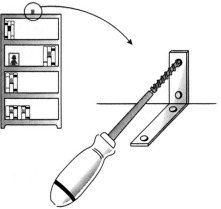

- Secure tall, free-standing furniture (bookcases, china cabinets, and shelving units) to the wall studs using L brackets, as shown in the diagram.

- Secure items such as televisions, stereos, computers, microwaves, and breakable collectibles with Velcro® brand fasteners or double-sided tape.

- Place heavy or breakable objects on lower shelves.

- Secure poisons, toxins, and solvents:

  o Move poisons, toxins, and solvents to a lower area, behind a guardrail, or in a locked cabinet.

  o Keep these items away from your water and food storage areas and out of the reach of children and pets.

  o Ideally, store all flammable liquids in an outside building, away from structures and vehicles. Spilled flammable liquids could easily cause a fire and destroy your home.

- Consider moving all framed pictures and mirrors away from high traffic areas, such as near beds, couches, hallways, and exits. Earthquakes have a tendency to knock pictures and mirrors off the walls, causing injury if they fall or leave sharp debris on the floor.

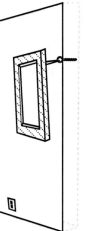

- Secure framed pictures and mirrors to the wall using eye hooks that penetrate the wall stud. Additionally, you could run a wire across the back of the picture or mirror and attach it to the eye hooks.

- Avoid hanging plants or other objects, such as lamps, near high traffic areas. Keep them away from windows to prevent them from swinging and breaking the glass. Secure these objects directly into a ceiling stud.

- Secure kitchen cabinets by installing a latch, as shown, to prevent cabinet doors from flying open.

- Wood-burning stoves or heaters are quite heavy. They can weigh over 220 pounds (100 kilograms) and can easily topple over from tremors, causing a fire or leaking smoke and other gases into the house. You may consider hiring a carpenter to secure your stove to ensure it will not move or topple over. If you choose do the work yourself, you might want to have it inspected afterwards.

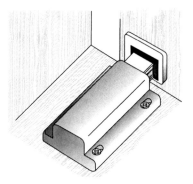

Prepare Your Home for a House Fire

Disasters, particularly earthquakes, may trigger a fire in your home. If you are in an earthquake risk zone, ensure you are prepared for fires. Have smoke detectors and fire extinguishers readily available in your home, garage, and any other buildings on the property.

This manual does not focus on fires that start within the home. For more detailed information about house fires, please visit the National Fire Protection Association at http://www.nfpa.org/.

Utilities

Preparing the home includes ensuring that every family member knows when and how to shut off utilities. This will help to reduce further risk to your family and protect your house and belongings from additional damage, such as fire, explosion, electrocution, and contamination from fuel.

Below are some guidelines for preparing your gas, electricity, portable generators, water supply, and propane tanks for disasters and extreme weather, as well as guidance on shutting off your utilities.

Gas

Natural gas is one of the safest gases; pipelines are built for maximum safety and minimal service interruption in an emergency. As well, leaks can be easily detected by a "rotten egg" or sulfurous odor, and a pressure regulator maintains a safe flow of gas to the gas meter and gas appliances.

In areas where earthquakes have occurred, gas delivery systems have withstood ground movement very well. Disruption of gas service and other underground utilities have typically been caused by the collapse of pipe-supporting structures such as bridges. Landslides can also damage gas delivery systems.

Despite the safety of natural gas, disasters and extreme weather can damage pipelines and cause leaks, which can result in explosions and fires. Here are some safety guidelines.

How to prepare and maintain your gas system

- Ensure there is a clear path to the main gas valve.

- Inspect the system to ensure it is free from debris, cracks, and damage.

- Keep the area around the main gas valve and gas appliances clear of combustible materials, including paper, paints, solvents, propane cylinders, barbecues, lawnmowers, and gas cans.

- Have a dedicated wrench easily accessible to close the valve.

- Ensure all family members know when, and how, to shut off the gas system.
- If you have appliances connected to natural gas, know where the shut-off valves are for these appliances. As well, do not store combustible materials (paper, paints, solvents, flammable chemicals) near the gas appliances.

When to shut off the gas

- If you smell gas, or hear a blowing or hissing noise, close the shut-off valve.

How to shut off the gas and respond to a leak

- In case of a leak, exit immediately, leaving the doors and windows open behind you.

- If you smell gas, do not smoke, light matches, use your cell phone, or operate electrical switches or any other ignition sources. The risk of a spark may cause a fire.

- If your safety is not compromised and you have time, turn off your gas meter outside using a dedicated wrench. Turn the switch one quarter turn to the right or left so that it is perpendicular to the gas line.

- Leave the area on foot; do not start your vehicle if it is close to the leak.

- Go to a nearby phone and call your natural gas company or 911.

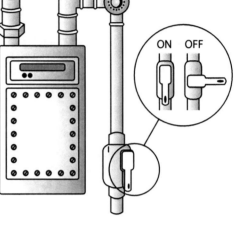

Note: Once the gas is turned off, only a qualified technician should turn it back on. The system should also be inspected and repaired before turning it back on.

Electricity

In case of a power outage, you may need to turn off the electricity to your house to reduce the risk of fire, especially if no one will be home when the power is restored. Electrical sparks have the potential to ignite if natural gas is leaking. As well, if there is a risk of flooding, shutting off the power will help reduce the risk of electrocution, as water conducts electricity. In the event you do not have time to turn off the main electrical power, leave an exterior light on to signify to local authorities (electric company, emergency response personnel) that you have power.

Note: Never go into a basement that is filled with water until the electricity has been completely disabled by authorities. See Step 6 for more details about electrical safety.

How to prepare and maintain your electrical system

- Ensure all family members are comfortable with shutting off the electricity.
- Label all switches in the electrical panel, including the one for the hot water heater.
- Inspect the panel to ensure easy access and that it is free from debris and damage.

When to turn off the electricity

- If you must evacuate your home.
- If power is disrupted during a disaster.
- If flooding is expected.
- If you have been advised to do so.

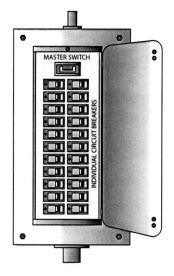

How to turn off and restore the electricity

- Turn off and unplug any sensitive equipment and appliances you do not need to use (computers, hot tubs, ovens).
- Discard any perishable food from your refrigerator and prop the door open.
- Turn off all the inside lights.
- Turn off the electricity as follows:
    - o From the electrical panel (typically in the basement or garage), turn off the individual breakers first, then the main breaker, so that when the power is restored, you can control how you turn on your appliances.
    - o Always shut off all the individual circuits before shutting off the main circuit breaker. Turn off the individual circuits by flipping them to the off position, and then flip the main breaker to the off position.
- When power is restored, turn on the main breaker first, then turn on individual circuits one at a time.

Note: See Step 7 for more information on what to do with the electricity after returning home from an evacuation.

Portable Generators

In the event of a power outage, portable generators can be very helpful. However, they can cause carbon monoxide poisoning, electric shock, and fire if they are not used properly. Here are some key safety guidelines:

- Do not connect the generator to your home's electrical system; rather, connect equipment directly to the generator using a three-pronged extension cord with a gauge heavy enough to accept the anticipated draw. Use caution not to exceed the maximum load.
- Never operate a generator indoors: the exhaust can cause carbon monoxide poisoning if inhaled. Carbon monoxide gas is particularly dangerous and hard to detect because it is colorless and may not have a noticeable odor.
- Position the generator outside, or in a well-ventilated garage or shed, where the exhaust cannot enter the house. Keep the generator protected from water, rain, and snow.
- Do not store fuel near the generator.
- Ensure the generator is completely shut down and cool before refueling.
- Maintain the generator in good repair and change the oil as per the manufacturer's guidelines.

You may also consider a solar-powered generator, which is safe to operate indoors and is handy for critical appliances and health care equipment. These generators come in various power outputs and are available at larger hardware stores.

Water Supply

Following a disaster, it is important to protect your water supply and prevent contamination. Disasters such as earthquakes and floods may damage the water distribution system and contaminate the water supply coming into your house. For these reasons, it's a good idea to close the main house water valve until health officials assure you there is no risk of contamination.

How to prepare and maintain your water heater and water system

- Secure your water heater as per the guidelines below.
- Check the main water valves and water connections regularly and replace them if necessary.
- Label the valves with tags for easy identification.
- Ensure all family members know how to shut off the water valves.
- If you cannot close the valves with your hands, place a tool close by.
- Include a wrench in your emergency kit as backup.

How to secure the water heater

A water heater can be very heavy when full (450 to 900 pounds; 200 to 400 kilograms). A sudden shake or motion from an earthquake can cause it to fall over and break the water pipes or the connectors. As well, if your water heater operates on natural gas, the movement can damage the gas line. Securing your water heater is fairly simple and in an emergency can protect your water supply from contamination, your tank from denting, and your home from water damage.

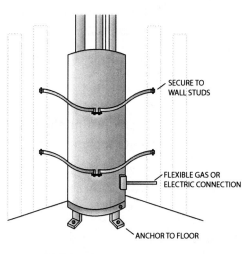

SECURE TO WALL STUDS

FLEXIBLE GAS OR ELECTRIC CONNECTION

ANCHOR TO FLOOR

- Secure the top and bottom of the heater to wood or metal studs so that it doesn't fall over during tremors. Metal straps work well.
- Anchor the bottom to the floor.
- Install a flexible gas or electric connection. If your water tank is powered by gas, have a licensed gas fitter install a flexible pipe to connect the gas supply once the tank is secured.

When to shut off the water valves

- In the event of a disaster, you will want to preserve as much water as possible. Shutting off the main house valve will prevent the water stored inside your hot water tank and toilet tanks from draining out of the house due to gravity.
- This valve may be in the basement, in the garage, or on the main floor. It controls the water inside the home and connects to the main street pipe that supplies water from the city. It will look like one of these:

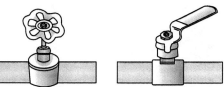

- You will want to shut off the main house valve ...
    - o If there is a water leak or broken pipe inside the house, or the water main outside is damaged.
    - o If there is no city water coming to your home. Shutting off the main house valve will prevent backflow of your water from the house to the city pipes. If there is backflow, your hot water tank and toilet tanks may drain out.
    - o If you have to evacuate your home.
- You may want to shut off the main street valve...
    - o If there is a water leak or break in the pipe between the house valve and the street valve.
    - o If there is no supply of city water coming to your home (to prevent backflow).

How to shut off the water

- For the main house shut-off valve:
    - o Turn the valve to the right to close the valve completely. This might be challenging, as it may be rusted open. You may need to use a wrench to close it completely.
    - o If you have an electric hot water tank, turn off the electric circuit for the tank to prevent the unit from burning out if you empty the tank in an emergency (when no other water is available).
- For the main street shut-off valve:
    - o The outside street shut-off valve is typically located in your front yard or in the driveway near your property line. It is usually in a concrete box and might not be completely centered within the box. You may also have to dig it out. Be aware that critters and spiders sometimes nest in the box.
    - o The street valve may be difficult to close and sometimes requires a special water key or wrench. You should check this valve in advance to ensure you can close it and have the necessary tool.

Note: Older constructions do not typically have a valve to prevent backflow. You may want to have one installed to protect the water inside your house.

Fuel, Oil, and Propane Tanks

Fuel and oil tanks can be dislodged or damaged by a flood, earthquake, tsunami, or extreme weather. This might cause a leak in the tank or fuel line and quickly pollute the soil and water system. These tanks should therefore be secured.

Propane tanks and other compressed gas tanks (e.g., oxygen) stored outside or in the garage might break free from their supporting legs and fall, creating a risk of fire or explosion. As well, the supply line might rupture, at which point escaping gas could cause a fire. Follow the instructions below to prevent damage to fuel and oil tanks and compressed gas cylinders.

How to prepare your fuel, oil, and propane tanks

- Regularly check tanks, legs, and support systems for damage or leaks. As well, check them after any tremors, high winds, or extreme weather.

- Keep the tanks free from tall or heavy objects that may fall and damage the tank or supply line.

- The tanks should be secured in place with chains (or similar hardware) at the top and bottom of the cylinder. Attach the chains to eyehooks securely screwed into a wall.

- You may consider hiring a professional to weld the bracing to the tank legs and install flexible connections to all supply lines running from the tank into the home. This will allow for some movement and prevent breaking the line. The professional may need to remove the gas from the tank and supply line before conducting any work.

- Ensure family members know how to detect a gas leak and how to turn off the supply line. Keep a wrench nearby.

- If you have a large tank, such as on farms, you might consider installing a seismic shut-off valve, which automatically turns off the gas during an earthquake.

When to shut off the tanks

- If the tanks are dislodged, damaged, or have a leak in the tank or the fuel line, shut them off.

How to shut off your propane tank

- Turning off the propane tank is a minor procedure—simply close the valve. You can turn the valve off and on any time without help from a technician, because your gas supply comes from a stand-alone tank.

Note: Be aware that propane is heavier than air and will pool in low areas, such as in the basement. Avoid these areas in the event of a propane leak.

Multi-Residential Complexes

If you live in multi-residential complex such as a townhouse, apartment, or condo, preparing for an emergency may take extra planning. These buildings and complexes are typically quite large, with many exits and shared spaces, numerous residents, and a number of people involved in emergency planning. It is important to understand the complex's emergency procedures, know your role if disaster strikes, and prepare your suite and belongings accordingly.

How to Prepare for a Disaster

Your townhouse, condo, or apartment complex should have a detailed emergency and evacuation plan. Although the property manager typically maintains the plan, it is important to participate in the planning process and understand your responsibility in case of emergency so that you can prepare accordingly.

Multi-residential complexes typically have sophisticated water, electricity, gas, heating, fire suppressant, and ventilation systems. Although the property managers are responsible for ensuring that these systems are maintained, each resident should have a basic knowledge of the building's emergency systems, and of the safety procedures in the event of an emergency.

Tips for Residents of Townhouses, Apartments, or Condominiums

Get to know the emergency plan

- Examine and understand the complex's emergency and evacuation plan.
- Discuss what would happen to the residents if your building were severely damaged by a disaster.

Participate in the plan

- Review the plan and provide suggestions for improvement. Everyone's input is vital to ensure the plan is effective and incorporates the needs of every resident.
- Inform the building management if you have special needs or require specific equipment to evacuate.
- Participate in an annual walk-through of the building to familiarize residents with its features. Residents can discuss the plan, note opportunities for improvement, and identify possible hazards in all areas (entrances, hallways, shared spaces, laundry room, parking). This is also a great opportunity to get to know the property managers and your neighbors. You might discuss one another's skills and special needs, as well as how to incorporate them into the plan. With this information, you might consider developing an emergency response team or at least nominating a few volunteers for specific tasks in the event of an emergency.

Prepare your suite

- Depending on the emergency and whether you live in a townhouse, condo, or apartment, your home will require different preparations. Follow the general house preparation guidelines and adjust accordingly.

Know about the utility systems

- Know about the building's backup utility systems, such as the auxiliary power, main shut-off valves, and power switches. Know who is responsible for shutting them off in an emergency and who has 24/7 emergency access.
- Know what utilities should be shut off in your suite. Know where the hot water tank, sinks, toilet, and gas valves are located and how to shut them off.

Know how to evacuate

- Review the evacuation plan and practice it regularly.
- Locate and familiarize yourself with all the emergency exits and stairwells.
    - o Do they have emergency lighting?
    - o How long will the lights stay on?
    - o Where are the emergency portable flashlights located?
    - o How will you exit the parking area?
    - o How will the garage door open if there is no power or it is damaged?
    - o How do you open the garage door manually?
    - o Is there another way to exit the parking area?

- Practice locating two emergency exits as you enter and leave the building from different routes.
- Remember that elevators should generally be avoided in a disaster.

Know about the general safety equipment

- Locate all the fire extinguishers, escape ladders, defibrillators, and spare wheelchairs—and know how to use them.
- Know the building superintendent's name and phone number.
- Know the contact names for each floor.
- In an emergency, know that the sprinkler systems and fire alarms may activate.

If your complex does not have a comprehensive emergency and evacuation plan with adequate disaster kits, encourage your building manager or owners to develop a system to ensure the residents are prepared for a disaster.

Suggestions for Managers and Owners of Multi-Residential Complexes

While this is not an exhaustive list, the following are some suggestions for property managers.

- Develop a comprehensive emergency and evacuation plan. This could prevent any liability issues following a disaster, as well as reduce damage and financial impact.
- Ensure all residents are familiar with the emergency plan.
- Run drills regularly to allow opportunities to improve the plan.
- Develop a warning system to alert residents of imminent danger.
- Make an emergency kit for residents.
- Develop response teams.
- Ensure adequate safety systems and backup utilities.
- Know how to respond to any emergency.
- Know how to safely evacuate all residents.
- Have supplies ready to repair any urgent damage.

Annual Maintenance

Maintain your kits at least once a year to ensure they are up to date and to familiarize yourself with the equipment. This is also a great opportunity to discuss your emergency plan with your family. As well, if your family situation has changed, such as with the addition of children or pets, adjust your kit accordingly. Pull a copy of this book out of your grab-and-go bag to review the checklists and skim through the content to keep what you've learned fresh in your mind.

Annual maintenance can be done anytime, as long as it gets done at least once a year. For ease of remembering, you might do the maintenance the week of your birthday, over the winter holidays, when school begins, or when you pay your taxes or home insurance. Set a reminder on a paper or electronic calendar so there's no chance of forgetting. Download a printable version of the annual maintenance checklist from my website, at www.authorkimfournier.com/resources/ (password *flashlight*).

The following are some annual maintenance tips:

- Visit a travel clinic to ensure your vaccinations are up to date.

- Ensure all your contact numbers, photos, and photocopies are up to date.

- Inspect all items for integrity and to ensure they are functioning well.

- Check the expiry dates of the food and medication.

- Rotate the water (you should be doing this every one to three months).

- Rotate canned foods. For optimal quality, any canned food should not be kept longer than one year, regardless of the expiry or best-before date.

- Examine dehydrated food. Discard if the package is compromised, if the food is expired, or if there is any sign of moisture or mold.

- Throw away any canned foods that are damaged, punctured, swollen, dented, or corroded.

- Revisit Steps 1 to 5 and explore how you might improve your plan, kits, inventory, location, food selection, tools, etc., and how you might better rotate the fresh water and food supply.

- Walk around your property and home with a focus on three main goals.

  o Reduce injury. Ask yourself: *What might cause injury in this area or room?*

  o Reduce loss or damage of property and possessions (including high-value items). Ask yourself: *How can I move or secure items to reduce loss or damage?*

  o Reduce loss or damage of essential items, including medication and comfort items. As you are examining each room, ask yourself: *What can I not live without in this area for a month?*

PART II: Respond

Congratulations! If you have completed Steps 1 to 5—or even parts thereof—take a moment to give yourself a well-deserved pat on the back. Disaster preparedness is accomplished one step at a time, something I've made easy with my 7 Step approach. Nevertheless, few of us get around to making it happen, even though I'll bet you and your family had some fun in the process. Well done!

With your important documents protected, your plans in place, your grab-and-go kits at the ready, and your house prepared, you are ready should a disaster strike your home or neighborhood. If that happens, however, you will also need to know how to respond. That's what we're going to tackle in Step 6: Respond to a Disaster. Let's get started.

STEP 6: Respond to a Disaster

Knowing how to respond to a disaster to ensure you minimize injury and loss takes a bit of planning. Although this manual focuses on an all-hazards approach and preparation strategies that support all disasters, a few emergencies will require unique response actions. This step includes these intricacies, along with other considerations such as whether you should stay in your home or evacuate, tips to shelter in your home, how to evacuate, suggestions on living in mass care shelters, and how to stay informed during a disaster.

> ### Step 6: Know how to respond to disasters.
> ### Follow the instructions in this section.

Shelter in Place or Evacuate

Depending on the type and severity of the emergency, you will need to decide quickly if you should shelter in place or evacuate. This decision is not always obvious. There are a number of factors to consider.

- Is your house safe to shelter in place?
- How much warning and preparation time do you have?
- What is the type of disaster?
- Do you have safe transportation?
- What are the road conditions?
- Is the community safe to evacuate?
- Are there any hazards on the way out of the city?
- Do you have a safe place to go?
- Does the community have temporary shelters?

Officials may make your decision of whether to evacuate easier by issuing one of three evacuation notices.

1. **Evacuation alert** warns of an impending danger and allows time for people to vacate voluntarily. Movement of vulnerable populations may commence at this time. This alert gives you valuable time to prepare: review your communication and emergency plan; assemble your emergency supplies; arrange a meeting place, transportation, and accommodations; and relocate pets if necessary.

2. **Evacuation order** occurs when the government declares a state of emergency and signifies an immediate and mandatory evacuation to ensure public safety. Be aware that these orders will be enforced. For your safety, it is best to comply. The authorities will have a comprehensive understanding of all the risks, some of which you may not be aware of.

3. **Rescinding of the evacuation notice** will be given once it is safe to return home—yeah! Details on returning home are covered in Step 7.

Shelter in Place

Tips to Shelter in Place

Shelter at home

- Choose the safest room, preferably a large room with a window, a door, and access to water and food, such as a master bedroom that is connected to a washroom.

- Ensure you can hear, or see, the warning systems in your neighborhood, such as lights or sirens.

- Contact your workplace, the children's school, and your immediate family members to check in on them and tell them where you are.

- Take turns listening to radio broadcasts and maintain a 24-hour safety watch.

- Try to contact the city to learn more about the situation and get further instructions.

- Review your emergency plan to guide your family's activities.

- Assemble your supplies and make a plan to ration the food and water based on the estimated time you will have to shelter. During extended periods of sheltering, you will need to manage water and food carefully. This may not be easy, but it is necessary.

- If you are not in urgent need of emergency services, place a sign in the front window stating "WE ARE OK." Accordingly, if you need help, place a sign in the window stating "WE NEED HELP!"

- Know that emergency responders cannot tend to every home. Be prepared to be self-sufficient and manage your own first aid needs, food and water supply, and urgent building repairs. See Step 4 for a suggested list of emergency supplies.

- Check in on neighbors frequently.

Shelter at work and school

If you and your children must shelter at work or at school, the same principles as those for sheltering at home apply. The key is to be familiar with your work and school emergency plans so that you know what to expect and how to prepare. Also, ensure there are adequate supplies at school, in your workplace, and in your vehicle to keep you going for a while.

In the event you are separated from your family, your communication plan will come in handy and reduce worry. Therefore, make sure it is updated annually and kept in several locations, including with a relative or friend.

- When sheltering away from home, follow the same instructions as for sheltering at home and adjust accordingly.

- Review the emergency plan and inspect supplies. The plan and the supplies should be checked regularly—supplies at work have a funny way of disappearing.

- Suggest volunteers for specific duties, such as first aid and radio watch.

- Understand that everyone handles stress differently—be kind and patient.

Shelter during a release of hazardous material in the air

During an accidental release of a hazardous airborne material (nonflammable), you may be instructed to shelter indoors until the material dissipates and it is safe to go outside. However, if the hazardous material is in the building where you are, you will likely be instructed to evacuate to a safer area. Likewise, if the substance is flammable, such as a gas, authorities will likely instruct you to evacuate to a different area.

If you must shelter indoors due to a hazardous substance:

- Go indoors immediately and stay there.
- Close all windows and doors.
- Turn off the furnace, air conditioners, fireplaces (gas or wood), clothes dryers, bathroom vents, and any exhaust systems or air exchangers.
- Do not smoke.
- If you feel the need, wear an N95 mask.
- Turn on the radio or television for updates.
- Consider sealing any cracks around the windows and doors with duct tape or towels.

Evacuate Safely

An evacuation may be required if your home is not safe and you must temporarily move to an alternate place. It is best to understand the process ahead of time and prepare accordingly.

Tips to Evacuate

Before a disaster

- Be prepared to evacuate quickly: have your kit ready and accessible, pre-plan the exits (doors, windows) and escape routes out of the house, and know your evacuation route out of the neighborhood. Ensure your evacuation plan is up to date.
- Have the necessary devices to escape safely, especially an axe in the attic and safety ladders on balconies and windows.
- Ensure everyone knows how to evacuate safely—practice in advance.
- Prepare your vehicle for evacuations. Always keep it in good repair and have at least a half a tank of gas. Keep in mind that gas stations may be closed and the pumps may not function, either because they are damaged or because there is no electricity to operate them. As well, gas companies may not be able to deliver fuel due to damaged roads or equipment.
- Become familiar with evacuation routes and alternative methods to leave the city.
- Always keep a bike handy in the event your vehicle is not serviceable. Keep a chain, lock and key, tire repair kit, and air pump on the bike. The roads may have sharp debris that can puncture the bike tires.
- Pets are generally not allowed in group lodgings. If you must leave your pets behind, prepare an emergency plan and supplies for them. This was covered under Preparing Your Pet for a Disaster on page 31.

During a disaster

- Before you leave, put on sturdy shoes and warm and waterproof clothing. As a backup plan, ensure there is spare clothing in your grab-and-go bags.

- If it is safe and you have time before you leave, do the following:
    - Turn off unnecessary appliances, such as hot tubs, air conditioners, and kitchen appliances.
    - Unplug sensitive electrical equipment, such as computers and televisions.
    - Remove perishable food from the fridge and dispose of it in an airtight garbage container.
    - Ensure your outside trashcans are tightly closed so they do not attract rodents.
    - Shut off the water, electricity, and fuel and propane tanks. If you've been instructed to do so, turn off the gas. See the Utilities section in Step 5, starting on page 40, for details on how and when to shut off the utilities.
    - Turn off all lights except for one exterior light. This will let city emergency workers know you still have electricity.
    - Lock the doors and windows.

- Help your neighbors evacuate.

- When you have time, call, text, or email a friend or family member to let them know where you will be going.

- When you are instructed that it is safe to return home, follow the instructions in Step 7: Return Home Safely.

Living in a Mass Care Shelter

If you have to evacuate and cannot stay with a friend, neighbor, or relative, staying in a mass care shelter may be an option if one is available near you. These shelters might be set up in public facilities such as schools, places of worship, shopping malls, and arenas. If they are available, remember that it takes a lot of coordination, logistics, and staff to set them up and communicate to the public where they are. It may take a few days to gather bedding, water, food, and basic sanitary facilities (toilets and areas to maintain personal hygiene) for a large population. Mass care shelters are not the most pleasant option and will require patience and as much self-sufficiency as you can manage.

These shelters involve living with many people in tight quarters, which can be unpleasant, loud, stressful, and unsanitary. Additionally, authorities may not be able to secure enough food, water, and cots for everyone. So be prepared—acknowledge what this experience might look like, discuss it with your family, and plan accordingly. Make sure you have enough water, food, medicine, and basic sanitary supplies in your evacuation kit for a long stay. Despite the unpleasantness, it is important that you stay in the shelter until authorities instruct you that it is safe to return home.

The following are suggestions on how to stay safe and comfortable during your stay in a mass care shelter:

- To avoid additional tension in this stressful situation, it is important to cooperate with shelter staff. If you can, help them.

- Ensure you register at the reception center inside or near the shelter. The reception center will assist with locating missing people, as well as coordinating vouchers for food, lodging, and basic necessities.

- Acknowledge that many people are living in a small space, all of whom have gone through the disaster: be mindful, understanding, and compassionate.

- Remember that your children and belongings are your responsibility. It is best to keep your valuables with you.

- Keep in mind that smoking, alcoholic beverages, and weapons are not permitted in the shelters.

- During extended stays, you will need to ration water, food, and supplies. Try not to focus on the lack of comforts or on the trauma of the situation; rather, remember that you are safe and alive.

- Don't forget to bring bedding, hygiene supplies, earplugs, flip flops for the communal showers, and entertainment and comfort items (music, books, games, teddy bear).

- Pets are generally not permitted in mass shelters, so pre-plan alternate shelter for them (see Step 4, Preparing Your Pet for a Disaster). Some communities may have emergency animal shelters.

Stay Informed During a Disaster

Emergency Notification Systems

If time permits, you may receive notice about an imminent disaster through various emergency notification systems, such as alarms, horns, radio or television broadcasts, phone, highway notifications, texts, and various websites. Get to know how your community intends to alert the public in case of an emergency.

Although there have been few studies about how animals can detect imminent danger, they can provide early warning. Animals have keen senses and instincts that may alert you to a disaster. If you notice your pets or neighborhood animals such as birds displaying unusual behavior, do not dismiss these clues, especially if your current circumstances indicate the potential for a disaster.

Communication During a Disaster

Disasters may affect communications in different ways, but it can be hard to predict exactly how, and for how long. Even if communication lines are working, they may be inconsistent and quickly overwhelmed by increased use. During a disaster, power, cable, Internet, and cell coverage may be limited, if not completely down. This will challenge how you access information and communicate with friends and family.

The following are suggestions on how to stay informed during a disaster:

- Do not tie up the lines; emergency response teams will need all the lines they can get.

- Using communication technologies like text messaging, email, or social media instead of telephones takes up less bandwidth and helps reduce network congestion after an emergency.

- If you must call someone, keep your call short and focus on urgent and pertinent information.

- Keep your emails and texts brief and avoid including attachments.

- Call 911 only if you are in immediate, life-threatening danger.

- Have alternative ways to charge your cell phone, such as extra batteries or a solar, hand-cranked, or vehicle charger.

- If you have a landline, keep at least one corded phone in the house.

- Power down your cell phone or turn it on airplane mode to extend battery life.

- Review your communication plan, how you intend to communicate with friends and family, and any predetermined locations where you plan to meet.

- Tune in to local radio, television, and Internet sites for updates and alerts. A list of stations should be included in your communication plan. You may also consider taping the list on the back of your radio.

Power Outages

Disasters often damage the power grid, leaving residents without electricity for days, sometime weeks. It is important to be prepared to live without electricity. The following are some suggestions to safely manage power outages:

- If your power goes out, determine if the outage has affected only your home or is neighborhood-wide.

- If the power outage has affected only your house, check your main electrical box and fuses. Replace fuses as required. If you are unsure how to fix the problem, contact an electrician.

- If the power outage is neighborhood-wide, check that your electricity service provider is aware of the outage by visiting their website or calling them. If not, inform your provider about the power outage and stay updated on the situation.

- Report any downed power lines to the power company, the fire department, and the police.

- Unplug all unnecessary appliances to avoid damage from a power surge when the power is restored.

- Turn off all the lights except for one to alert you when the power is restored.

- Use surge protectors with sensitive equipment to protect it from power surges when the electricity returns.

- Turn your thermostats down so that they are not constantly calling for heat.

- Natural gas furnaces typically need electricity to function and therefore do not have to be turned off during a power failure. They should operate normally once the power is restored. However, some natural gas fireplace burners generally function without electricity. Either way, you may consider purchasing a backup heat source.

- To stay warm, wear warm clothes (winter hats help a lot) and keep the windows closed. Consider closing off one room and heating that space only; choose a sunny, south-facing room. You can also set up a tent in one room.

- Remember, the food in your fridge and freezer may spoil if the power is out for a while. Try to keep the fridge and freezer doors closed to trap the cold inside.

- Conserve water in the event that city pumping stations are not able to provide water to your home.

- Use portable generators safely (see page 42).

- Keep informed of the situation: turn on a battery-powered or wind-up radio and tune in to the local station.

- Once the electricity has been restored and the power is stabilized, turn the heat, lights, and appliances back on. You may have to reignite the pilot light on certain appliances. Step 7: Return Home Safely has more details about turning your electricity, gas, and water back on.

Response Actions for Unique Hazards

Extreme Weather

We are experiencing more extreme weather than ever before, including heat waves, severe storms, ice storms, high winds, snow, hail, rain, drought, and thunderstorms. Regardless of where you live, you should be prepared for the possibility of extreme weather, which creates a number of risks to be aware of.

In Step 5, you took steps to prepare your house from wildfires, floods and freeze-ups, and earthquakes. Those measures will prove invaluable in case of extreme weather. In addition, during extreme weather, follow these guidelines:

- Cold weather can lead to hypothermia and frostbite; try to stay indoors or take shelter to stay warm.

- Hot weather can lead to heat stress, exhaustion, and dehydration. Protect yourself by staying cool and hydrated.

- Avoid driving. Slippery, icy roads and reduced visibility increase the likelihood of car accidents. Storms can lead to dangerous debris flying through the air or blocking the roadways.

- Stay indoors to avoid the following risks:

    o Injury or electrocution from downed power lines.

    o Injury from being struck by fast-moving objects (branches, loose garbage cans, etc.).

    o Flooding from storm surges or fast melting snow.

    o Injury from falling on slippery or icy sidewalks and stairs, or from ice dropping from branches or power lines.

Severe storms, ice build-up, wind, and lightening can damage trees or power lines. Follow these tips if you come across a downed power line or exposed power cable:

- If you must go outside, stay thirty feet (ten meters) away from trees and loose or fallen power lines.

- Assume all power lines are energized.

- Look around and above to see if there are any other damaged lines, and stay back at least thirty feet (ten meters).

- If you are driving and your vehicle makes contact with an energized power line, stay inside until help arrives. If you must get out due to fire, jump out with your feet together and do not make contact with any other parts of the vehicle on the way out. Once out, slowly shuffle away, keeping both feet close together.

- Do not touch a person who has received an electrical shock, especially if the person is still in contact with the source. The power may travel into your body. Call 911.

Tsunamis

While earthquakes and tsunamis are difficult to predict, authorities suggest that those who live within a thirteen-foot (four-meter) elevation of the normal highest tide should plan for tsunamis. Know the risks and have your emergency kit ready.

Visit your local disaster manager; this person will likely have a hazard map available for your area that includes the estimated tsunami arrival time, wave height, and safe evacuation area. Know the signs of a tsunami:

- A sudden rise or fall of the ocean level, a loud noise coming from the ocean, and/or shaking ground.
- Waves may continue to come following the initial wave. Typically, there are several waves separated by minutes or by up to an hour.
- Generally, the first wave is preceded by a recession of the sea, and the second and third waves are the largest.

Tsunami alerts may or may not be issued. Nonetheless, you should be familiar with the relevant terms if you are in an area that may be prone to tsunamis. There are three main types of alerts:

1. **Tsunami watch** is the least serious alert, issued when a danger is known. It serves to encourage people to stay informed for updates. If a tsunami watch is issued, assemble your emergency kit, stay informed, and be prepared to evacuate quickly. Check with your neighbors to ensure they are aware of the alert.
2. **Tsunami advisory** is the second highest alert and serves to notify people to stay away from the shore.
3. **Tsunami warning** is the most serious alert, indicating that a tsunami is likely and evacuation may be suggested.

Regardless of the warnings, if you are in a tsunami zone and feel the ground shaking for twenty seconds or more, you should seek higher ground. Likewise, if you are on the beach and notice a sudden rise or fall of the ocean level, this is an important sign that a tsunami may be approaching. Get away from the shore and make for higher ground immediately.

If a tsunami alert is issued, or if you think a tsunami may be approaching, take these steps:

- Double-check your emergency kit and take it with you if you need to evacuate.
- If you have time, check on your neighbors to ensure they know of the tsunami alert.
- Listen to the local radio station for updates.
- Seek higher ground if you are less than thirteen feet (four meters) above sea level. If you live along the coast within a tsunami zone, you will want to seek ground sixty-five feet (twenty meters) or higher.
- If you are on a boat, leave the harbor and go to open water.
- Expect several waves over many hours following the initial one. The second and third waves may be bigger than the initial wave.
- Return home only when officials indicate it is safe to do so.

Wildfires

Wildfires can greatly disrupt a community, causing extensive damage and destroying property. They often begin unnoticed, are unpredictable, and spread quickly, allowing little time to evacuate. Preparing for such a devastating disaster is paramount, which you did as part of Step 5: Prepare the House.

In addition to those instructions, the following are guidelines for responding to fires:

- If the fire is in your home,
    o Get out of the house quickly.
    o If you have time and it is safe, close the doors and windows on your way out.
    o If you are trapped inside a room, hang a sheet in an open window to let firefighters know where you are, then stay low to the floor and close the door.

- If there is a wildfire near your home,
    o Prepare your home and family for evacuation. If you are put on alert of impending danger, review your emergency plan, assemble your emergency supplies, arrange transportation and accommodations, and relocate pets if necessary.
    o If you must evacuate, do so safely (see Evacuate Safely above).

Communicable Diseases and Pandemics

Pandemics can spread very quickly through direct or indirect contact, significantly affecting people in every industry and seriously disrupting all aspects of a community. Following a disaster, communicable disease can occur in various ways:

- Insects and animals can cause vector-borne diseases such as West Nile virus, Lyme disease, and rabies.

- Consumption of food or water contaminated with bacteria such as E. coli, salmonella, and giardia may cause diseases.

- Direct or indirect contact with bodily fluids may cause diseases such as measles, mumps, rhinovirus (common cold), tuberculosis, SARS, and influenza (e.g., H1N1).

Whatever the disease, prevention is your best defense:

- Wash your hands frequently with soap and water. This is the most effective method to clean your hands. In the absence of soap and water, use an alcohol-based hand sanitizer. Sanitizers that do not contain alcohol are not effective at destroying pathogens that cause infectious diseases.

- Cover your mouth and nose with a tissue when you sneeze or cough. Discard the tissue after use.

- If you don't have a tissue, cough into your sleeve. Wash your hands afterwards. Do not use handkerchiefs.

- Avoid sharing personal items that come into contact with the skin, such as towels, cups, lipstick, water bottles, and hair brushes.

- Get lots of rest, and consume lots of water and nutritious food.

- Practice safe handling and storage of food and water.

- Keep your vaccinations up to date. Get the flu shot.

- If you are sick, stay home.

- Avoid unnecessary contact with people who may be contagious. Stay three feet (one metre) away from people who are ill.

- Avoid touching surfaces that are particularly unclean and pose a high risk for cross-contamination, such as door knobs, hand rails, and counter tops.

- Avoid public gatherings.

- Keep updated about public health emergencies from credible sources, such as public health officers and local health authority representatives on television and radio stations.

- Do not trust information from unreliable social media or Internet sources. False information and rumours spread quickly during public health emergencies and can negatively affect your health and increase the spread of the disease.

- Follow the directions of health officials regarding any infectious disease outbreaks. For example, you may be instructed to boil your water.

- Use the gloves, face masks, hand soap, and alcohol-based hand sanitizer you included in your emergency kit as needed to protect yourself and your family. Use your supply of household chlorine bleach to disinfect surfaces and equipment, or to sanitize water.

Earthquakes

Earthquakes can occur at any time, and can happen suddenly and without warning. Prepare your family, home, vehicle, and workspace for an earthquake as discussed in Step 5 and follow the information below on how to respond.

Wherever you are when an earthquake starts, drop, cover, and hold on:

1. **Drop** to your hands and knees before the earthquake knocks you down. This position protects you from falling but still allows you to move around. If possible, quickly get under a nearby table, desk, or heavy furniture. If this is not possible, crouch down against an interior or corner wall. Avoid doorways and stay away from windows and heavy, loose objects. If you are in a wheelchair, drop your body forward and lock the wheels.

2. **Cover** your head and neck with your arms to protect yourself from falling objects.

3. **Hold on** to something sturdy, preferably with overhead protection, until the shaking stops. Be prepared to move with this object if the shaking shifts you around.

4. Remain in a protected place for sixty seconds after the shaking stops. Be prepared for aftershocks soon after the first quake.

The following table details how to respond to an earthquake in various locations. However, the basic actions are the same: drop, cover, hold on, and remain in place until it is safe. When you do move, be aware of potential hazards around you.

Location	How to Respond to an Earthquake
Home	Quickly get under a nearby table, desk, or piece of heavy furniture. If this is not possible, crouch down against an interior or corner wall. Avoid doorways and stay away from windows and heavy, loose objects.
Vehicle	Pull over, stop the vehicle, and apply the parking break. Avoid overpasses, bridges, and power lines. Assume all power lines are live. If a power line falls on your vehicle, you are at risk of an electrical shock. Stay in your vehicle until trained responders remove the wire.
Boat, ferry	Stay calm and follow the crew's direction.
Highrise	Drop, cover, and hold on. Face away from windows. If it is safe, move away from falling objects (bookcases, mirrors) and exterior walls. If you must exit the building, wait sixty seconds after the tremors stop and exit by a safe path (always know several exit routes). Do not use the elevators.
Wheelchair	Protect your head and neck with your arms or whatever is available (pillow, book). Lock the wheels and remain seated until sixty seconds after the shaking stops.
Outdoors	If it is safe to do so, move to a clear area away from hazards and drop to the ground. Avoid anything that could collapse, such as power lines, trees, bridges, overpasses, underpasses, buildings, traffic, and streetlights. A bus bench or picnic table can provide protection.
Bus, train, stadium, theater	Stay seated and crouch down. Protect your head and neck with your arms. Do not leave until sixty seconds after the shaking has stopped. Exit calmly—do not run.
Elevator	If you're in an elevator when an earthquake starts, hit all of the floor buttons and get out of the elevator as soon as you can.
Bed	Stay in bed, hold on, and protect your head with a pillow. You are less likely to be injured there and could avoid broken glass on the floor. Have socks, a sturdy pair of shoes, and a flashlight ready under your bed in case you need to evacuate.
Store	Drop, cover, and hold on. Avoid windows, skylights, and heavy objects that may fall. A shopping cart can provide some protection. Do not run for the exits or use an elevator.
In a coastal area	There may be a risk of a tsunami following the earthquake. Drop, cover, and hold on until the tremors stop. If the shaking lasts twenty seconds or more, move to higher ground.

After the shaking stops, stay calm and move cautiously, as there may be falling debris.

- If you are outdoors, stay clear of buildings, wires, and falling debris or unstable objects.
- Put out any small fires if it is safe to do so. Be aware that fire alarms and sprinkler systems may go off.
- When it is safe, secure loose items as required.
- If you are injured, treat yourself first before assisting others.
- If you are indoors and it is safe, stay there.

- If you are at home, determine if it is safe to stay based on structural damage, electrical and gas safety, and other hazards. Assess the situation and implement your family emergency plan.

- If you are evacuating, bring your grab-and-go bag with you.

- Follow the instructions on when and how to shut off your utilities (see Step 5, page 40).

- Connect with your family as per your communication plan.

- Check in on your neighbors.

Earthquake misconceptions debunked

Three major misconceptions persist about what to do in case of an earthquake. Make sure you avoid all three.

1. **DO NOT shelter in doorways:** A number of years ago, following a California earthquake, a photo circulated on the Internet of a collapsed home with the doorframe intact. From this photo, viewers came to believe that a doorway is the safest place to shelter during an earthquake. This is false.

   We now know that doorways are no stronger than any other part of the house and do not provide protection from falling or flying objects. Do not rely on doorways for protection. In fact, you are more exposed and at risk in doorways; you are much safer under a table. During an earthquake, get under a sturdy piece of furniture and hold on. This will help shelter you from falling objects that could injure you.

2. **DO NOT use the "triangle of life":** In recent years, information and photos have circulated on the Internet recommending people take cover in areas that may provide a triangle of protection. However, this is potentially life-threatening advice, and the source has since been discredited by experts in the field.

   The most effective action is to take cover immediately under a nearby table, desk, or piece of heavy furniture. If no furniture is around you, crouch down against an interior or corner wall. Avoid doorways and stay away from windows and heavy loose objects.

3. **DO NOT run outside:** People instinctively want to go outside when they feel tremors. However, running outside during an earthquake is dangerous: the ground is unstable, there may be debris, and you are at risk from falling objects such as glass, bricks, and electrical wires. You are at greater risk of injuring yourself outside. If you are inside when the shaking begins, stay there until it is safe to move around.

Floods

Floods occur frequently and can happen at any time of year. See the table below for a few tips on how to respond during a flood based on where you might be when one occurs:

Location	Steps to Take
At home	• Gas, live electrical wires, and floodwater can be a dangerous combination. If flooding occurs in your home, turn off the main power switch to your utilities and close the main gas valve, if advised to do so by authorities. As well, turn off propane valves at the tanks. • If water levels are high enough to cover your natural gas meter, call the gas company to come to check your meters before using any equipment. • If you have time and it is safe, move your valuables and electrical appliances upstairs, or away from the floodwater. Remain on the upper floor, keep updated as to the situation, and follow instructions from local authorities.
Outdoors	• Stay away from streams and rivers, and head for an elevated area. Floods can come quickly and moving waters can be very powerful and sweep you away. • Avoid floodwater and buildings surrounded by floodwater. The water may contain hazardous and sharp debris, as well as be contaminated by fuel, sewage, and chemicals. Water may also be electrically charged from underground or fallen power lines.
Anywhere	• If you have come into contact with floodwater, wash your hands thoroughly.

PART III: Return Home

As Dorothy so famously said in *The Wizard of Oz*, "There's no place like home." If you have been evacuated, it's likely that returning home will be the first thing on your mind for you and your family. However, there are a number of important safety tasks to consider before returning home, which I cover here in Step 7: Return Home Safely.

Returning home after a disaster can be very emotional and overwhelming, particularly if the loss or damage is great. It's important to be psychologically prepared for this possibility. In addition to the information I cover in Step 7, the information provided in Disaster Mental Wellness (starting on page 93) is important supplementary reading on psychosocial responses to disasters. Knowing how you and others are likely to respond in the event of a disaster can help prevent additional suffering.

Equally important is to prioritize the work that needs to be done to make your home safe and liveable again. The methodical approach I recommend puts safety first and helps protect against exhaustion. Additionally, I've provided general tips on what to do after a flood, including electrical safety and clean-up procedures. Thankfully, the work awaiting you should be much less devastating after having implemented the steps outlined in Part I. It is my sincere hope that before you know it, life will be back to normal for you—and "normal," of course, includes being well prepared should disaster strike again.

Step 7: Return Home Safely

Returning home after a disaster will require some preparation, both physically and mentally. It will feel wonderful to return home after such a traumatic event, but also sad to see your home and belongings damaged. Either way, do not rush the process. There are a few things to do to avoid a dangerous situation and ensure a safe return.

> ### *Step 7: Return home safely.*
> ### *Follow the instructions in this section.*

General Considerations

Tasks Before Returning Home

Before you head home, consider the following questions:

- Have the authorities cleared the neighborhood and confirmed that it is safe for you to return to your house?

- Are the roads safe to drive on?

- Are there any post-disaster risks, such as residual tremors, landslides, fires, or chemical spills?

- Are there any electrical hazards or hazardous materials in the area?

- Have the utilities been restored (electricity, gas, water)?

- Have the health authorities confirmed that the water is potable?

- Is your house safe to live in? Is there any major structural damage?

Tasks Upon Returning Home

- Before approaching your property or entering your home, check for hidden hazards. Examine your yard, the exterior of the house, the roof, and the windows.

- When you first return home, use a flashlight to examine the inside of the house carefully. In case there is a gas leak, avoid using candles, matches, or gas-powered lanterns to prevent sparks.

- Check gas, oil, and electrical systems.

- When the power is restored, turn on the main breaker first, then turn on individual circuits one at a time.

- If you turned off the main gas valve (at the meter), only a qualified technician (gas-fitter) should turn it back on. The system should also be inspected and repaired before turning it back on.

- Does the drinking water smell and look clean? If not, run the water to flush out the system before consuming any water. If in doubt, have your water tested for potability. Commercial laboratories generally provide water testing.

- If you have shut off the water, turn on the main water valve and the valve for the hot water heater. Open all the taps and allow fresh water to completely flush out the system until the water is clear with no odor.

- Check the sewage system to make sure there is no damage or leaking.
- If any wild animals have entered your home, do not attempt to approach them. They may be scared and dangerous. Never corner them, try to rescue them yourself, or shelter them in your home. Call your local animal control office.
- Do not try to move dead animals. The carcasses may be a health risk. Contact the local health department for instructions.
- If a wild animal bites you, seek immediate medical attention.
- Explore your options for financial help to repair your home. Make an insurance claim if you have coverage.
- Examine damaged items closely and take pictures for your insurance claim. It is important to inspect appliances; damaged appliances may not be safe to use. You may want to keep a waterproof disposable camera in your grab-and-go bag. Remember to keep all your electronics in waterproof, resealable bags.
- Examine your food, medicine, toiletries, and cosmetics closely for any moisture, pests, damage from rodents, or signs of contamination. If there is any unusual odor, discard it. If in doubt, throw it out!
- Sort out contents to be repaired or discarded; remove debris and clean any soiled belongings.
- If there was flooding, be conscious of the electrical safety measures and proper clean-up procedures I detail below.

Avoiding Exhaustion

Once you have settled back into your home, do not underestimate the trauma of the experience. Take time to rest, eat well, and have fun, and try to return to your normal routine sooner rather than later—this will be very grounding. Attend community debriefing sessions to help process the trauma.

There will be a lot to do; prioritize to avoid exhaustion. Keep a manageable schedule to repair and recover from damage.

Children will be particularly affected by disasters, and the recovery period can be quite an unstable time for them. They may not understand what has happened or why. Encourage them to share their feelings, draw, and tell stories. Be honest about what has happened. Include children in as many tasks as possible and encourage them to participate in the clean-up. Hug them often; human touch is very healing.

Flood Safety and Clean-Up

A flood can be very destructive. The water can get into everything, including the electrical system, causing a major safety hazard. Floodwaters also carry many pathogens that can contaminate drinking water, food, and belongings. Special procedures are required to prevent electrocution and to clean up after a flood.

After a Flood

- Return home only when officials announce it is safe. You may not be able to return home until authorities can secure safe water, arrange utilities inspection and restoration, and provide serviceable sewage systems.
- If flooding remains in or around a building, do not enter until the power company has disconnected the electricity.

- Do not enter a flood-damaged home or building if there is any risk of injury, illness, or electrocution (see further details in the next section). Likewise, do not enter if you smell gas or propane.

- If the home is free from floodwater, ensure the main power switch at the breaker box is off. If the breaker box is moist, do not touch it with your bare hands; rather, use a dry stick to close the switch.

- If it is safe to approach the home, assess the situation and conduct a perimeter check. Look outside for any damage, such as loose power lines, damaged gas lines, or hazardous or sharp debris. Before entering, examine the home for structural damage, such as foundation cracks. If there are substantial cracks, contact a building inspector or structural engineer for advice.

- Before using any large appliances and heating or cooling systems that have come into contact with floodwater, have them inspected by a technician. Some parts (wires, filters, fuses, valves) may need to be replaced to ensure their safety.

- Your drinking water may be contaminated. Listen for health authority boil water advisories and instructions on how to disinfect your water.

- If your well was flooded, it is likely contaminated and will need to be disinfected. Check with your local environmental health officer for instructions.

- Septic tanks will not normally be affected by flooding, but the septic field may be damaged. You may want to consult with a technical expert before using the septic field after a flood.

- Take an inventory and pictures of the damage, then move your belongings out of the floodwater to prepare for clean-up as per the instructions below.

- Contact your insurance company to initiate a damage claim.

Electrical Safety after a Flood

Although you may be tempted to return home and start the clean-up, do not rush into your house or basement after a flood without carefully evaluating the situation. If your house has been severely damaged by floodwater, there may be a serious risk of electrocution. The following are a few suggestions to help keep you safe:

- Never enter a flooded, or previously flooded, basement until the utility company, fire department, or an electrician has disconnected the electrical meter from its socket. This is the only way to completely disconnect from the grid and prevent electrocution. Keep in mind that even though you may not have power, there is still a risk of electrocution in a flooded basement from faulty circuit breakers and electricity back-feeding into the grid. The only way to guarantee safety is to disconnect from the meter.

- Before reconnecting to the grid, have all wiring that has been flooded inspected to prevent fires or dangerous short circuits. Your local utility company will be able to provide additional information.

- If there is a chance that flooding may have stressed gas piping, have the gas company check your meter before you use your gas system.

- Once the building is pumped, electrical equipment may need to be replaced, such as wires, cables, circuit panels and breakers, fuse boxes, fuses, switches, receptacles, lights, heaters, air conditioners, furnaces, and metal components of a home's electrical system.

Flood Clean-Up

Floodwater can contain dangerous pollutants such as raw sewage, agricultural waste, and harmful chemicals that can contaminate local waterways. Diseases can spread through contaminated water and sewage backup in several ways. Heavy rain can impact water treatment plants and sewage systems, and contaminate drinking water. As well, sewer systems can back up into household plumbing and increase the risk of contaminating food and water.

Diligent clean-up following floods is essential to reduce your risk of disease. Once your home is safe to enter, clean-up begins. Listed below are a few general safety tips regarding what to discard and how to clean contaminated items. For more detailed information, contact your local environmental health department.

General safety tips

- Items and surfaces that come into contact with floodwater should be carefully cleaned and dried as soon as possible to avoid mold growth.

- Wear protective clothing, such as rubber boots and gloves.

- Wear a face mask (N95) when cleaning areas that may have mold, asbestos, heavy dust, or dried animal excrement.

- Avoid skin contact with items that have come into contact with floodwater.

- Do not use contaminated water for washing, cleaning, drinking, or food preparation.

- Wash contaminated items and contact surfaces (including walls and floors) with a solution of one part household bleach to ten parts water. This will disinfect and remove any harmful pathogens, including mold.

- Mold is common after flooding and thrives in moist areas. Dry all items, dehumidify your home, and ventilate (open windows) as best as you can. The mold should eventually disappear.

- When cleaning up chemical spills or items soaked in chemicals, dispose of the soiled rags.

- Dry the rooms out, then clean the walls, ceilings, and floors.

- Dry out electrical circuits and boxes.

- Replace drywall and insulation, going two feet (sixty centimeters) above the water line.

- Remove flooring, carpets, and underlay that have been soaked with floodwater or sewage. Clean and dry the subfloor thoroughly before applying new flooring.

- Maintain diligent hand hygiene to prevent the spread of disease. Wash your hands frequently with soap and clean water, particularly before preparing or eating food, after toilet use, and after flood clean-up.

- If your basement is flooded, pump it out gradually—no more than about one third of the water per day—to prevent the water-logged walls from collapsing.

- If standing water remains in your basement and you are not able to dry it out, pour one half gallon (one to two liters) of chlorine bleach into the water and mix evenly. Repeat every four days if standing water remains.

Food

Food contamination and subsequent illness can occur following flooding and power outages due to food coming into contact with contaminated water and inadequate refrigeration, respectively. Here are a few tips to avoid illness from contaminated food:

- Discard food or other items that have come into contact with floodwater. If cans are intact (no rust or punctures), wash the outside of the can thoroughly and scrub the rim with a brush.

- Discard items from the fridge if you lost power for more than four hours. Discard items from the freezer if you lost power for forty-eight hours or more (twenty-four hours if your freezer was only half full).

- If in doubt, throw it out!

Summary

Step 1: Safeguard important documents

Keep copies of documents that may be destroyed or unavailable after a disaster in a separate safe location that protects the sensitivity of the information, such as at home in a fireproof safety box, in a safety deposit box at a bank, or with an out-of-town contact you trust.

Step 2: Create a communication and reunion plan

Discuss the plan with your family to make sure everyone is comfortable with it. You may also want to test it out with a practice emergency scenario to confirm that it works well. Every family member should have a copy of the plan. You may want to keep a copy in your purse, at the office, in the kids' backpacks, in your grab-and-go bag, and with your out-of-town contact.

Step 3: Create a shelter and evacuation plan

Discuss the plan with your family and practice it to make sure everyone is comfortable with different escape routes out of the house and evacuation routes out of the neighborhood. Remove any obstacles or hazards in the house that may restrict safe evacuation. Post the shelter and evacuation plan in a memorable part of the house so it is accessible in an emergency and everyone can review it from time to time, such as on the inside of a cabinet door. Make a copy to put in your grab-and-go bag.

Step 4: Assemble emergency supplies

Assemble the following emergency supplies:

- a grab-and-go bag
- stockpile of items for the home
- work, school, and vehicle kits
- a kit for your pet, if you have one

Include essential supplies for food, water, warmth, safety, hygiene and sanitation, mental wellness, and first aid. Supplies should last for at least seven days—aim for three weeks.

Step 5: Prepare the house

Prepare your home and property for disasters. Prepare your utilities and ensure your family knows when, and how, to shut off the gas, electricity, and water systems. Depending on the hazards in your area, you may also want to take extra steps to prepare for wildfires, floods, and earthquakes.

Step 6: Respond to a disaster

Be prepared to shelter in place or evacuate. Know how to stay informed about the situation and how to effectively communicate with loved ones after a disaster. Be prepared to live without electricity. Know the intricacies of how to respond to a wildfire, extreme weather, an earthquake, a tsunami, and a flood.

Step 7: Return home safely

Be familiar with the tasks required when you return home to avoid injury or illness.

You are totally prepared!

Additional Information

This final section is chock full of extra information you might find useful in preparing for emergencies.

Disaster 101 77

Hygiene and Sanitation..83

Collecting Water from Your Hot Water Tank...87

Collecting and Disinfecting Water..89

Disaster Mental Wellness...93

Special Needs 97

Disaster Insurance 101

Shopping List 105

Regardless of where you live or what your local hazards may be, I suggest you peruse Disaster 101, my primer on what happens during a disaster. It includes information on the specific hazards of extreme weather, wildfires, floods, earthquakes, tsunamis, and communicable diseases.

Communicable diseases can also become a secondary event, spreading quickly following a disaster. Good hygiene and sanitation go a long way in stopping the spread of those diseases. Knowing what to do may spare your family extra suffering, so I've included a number of tips for protecting against them.

During a disaster, potable water becomes a precious commodity. I cannot stress enough the importance of knowing how to collect and disinfect water if your supply runs out—it could save your life. I therefore outline instructions for water collection and sanitation in this section as well.

This section also covers disaster mental wellness and special needs considerations, such as family members requiring medical equipment or supplies.

Last but not least, I have included information on the all-important issue of disaster insurance. Regardless of the coverage you have now, I urge you to give this section a read. On top of a natural disaster, your family should not also have to suffer through a corresponding financial disaster.

Finally, starting on page 105, I've compiled a comprehensive shopping list to make it easy to get your supplies purchased and organized. If you prefer, you can also download this list from my website, www.authorkimfournier.com/resources/ (password *flashlight*). Happy shopping!

Disaster 101

Disasters can be complex and may require an interdisciplinary management and response process. Each event is unique and dynamic, as are its impacts. This section explains the impacts of disasters and how they are managed. As well, I describe six common hazards that could turn disastrous.

The Disaster Management Process

Disasters affect every part of society. Since most regions have a diverse geography, mixed demographics, and both urban and rural communities, disaster management is a complex process requiring cooperation among different sectors. Managing disasters consists primarily of three phases: preparedness, response, and recovery. There are numerous players involved in every phase, including planners; all levels of government; small businesses; health care organizations; emergency services (fire, ambulance, police); search and rescue operations; environmental and climate change specialists; shelter, food, water, and waste management providers; utility, transportation, and construction companies; and volunteer and relief organizations.

Despite the popular belief that the government and these various organizations will handle disasters, these groups have limited resources. They are not able to prepare, respond to, or help recover every family and business. For this reason, effective disaster management also relies on individual and family preparedness.

Emergency preparedness and risk reduction solutions are everyone's responsibility and should be integrated into every area of society—every family, community, school, business, and cultural organization. By being more self-sufficient and less dependent on external resources that may never come, we can reduce our losses and protect our own families and livelihoods.

Disaster Response

When a disaster hits, local emergency management crews respond immediately. If they need additional assistance, they request help from state or provincial resources. If the emergency escalates, the support of federal and nongovernmental organizations' support is mobilized. Depending on the situation, additional assistance could include national defense, border security, police, firefighting, search and rescue, environmental and health protection, and various private resources such as utilities, water and sanitation, and restoration. National or international resources are engaged and scaled up or down as required. This process is dynamic and tailored to each situation.

Effects of a Disaster

Disasters disrupt the normal functioning of our communities, including important infrastructure and amenities such as emergency services, hospitals, food, utilities, roads, rail, communications, water, utilities, lodgings, government support, and critical commercial businesses. The damage incurred depends, of course, on the type of disaster. Effects will vary in intensity and duration, and may have short- or long-term implications. Some events will be localized (e.g., hazardous material spills), while others may be more complex, spread over a larger area, and affect a wider array of services (e.g., earthquakes, floods, wildfires).

To make matters worse, most infrastructure in North America—including bridges, communication towers, and transportation facilities—was built prior to current disaster legislation, upgraded safety codes, and advanced engineering standards, all of which are designed to improve protection of the public.

In addition, many shelters and gathering places that would normally be used after a disaster—such as schools, hockey arenas, and hospitals—were also built prior to these standards and therefore have the potential to sustain extensive damage during a disaster.

Many public buildings are slowly undergoing upgrades to meet the current standards, but this will take time, money, and resources. Meanwhile, disasters continue to occur more quickly than the upgrades, highlighting the importance of individual preparedness and self-sufficiency to reduce the need for external assistance.

It is important to understand the many ways disasters might directly or indirectly affect your family and community. The biggest effects on people are death, injury, illness, and evacuation. The following table summarizes other potential outcomes that might occur from any type of disaster.

Entity	Types of damage	Other potential outcomes
Houses	Damage to structural integrity, windows, potable water and sewage systems, wells, and appliances.	Fires; water damage; shortage of furnace oil, electricity, or water; falling of unsecured items.
Vehicle	Dents, broken windows, flat tires.	Lack of gas.
Buildings	Damage to elevators, staircases, and windows; shattered glass; collapsed or flooded basement or underground parking.	Triggered alarms and sprinkler systems. Tall buildings and unreinforced masonry structures are particularly at risk during earthquakes.
Community infrastructure	Damage to bridges, roads, highways, air and sea ports, railways, water and sewage treatment plants, and dams.	Movement of goods and services, including emergency services, may be impeded. Gas stations and tellers may be challenged, and there may be a limited supply of food, medications, and pharmacy and hardware items (for home repair). Garbage disposal may be reduced.
Utilities	Damage to oil and gas pipelines, electrical grids, and sewage systems.	Loss of power, potable water, or sewage treatment. Limited or no heat or air conditioning. Hygiene and sanitation may be challenged. Fridges may not work, compromising food quality.
Communications	Damage to cellular, land line, or Internet connections.	Difficulties communicating, connecting with friends and family, and getting up-to date information about the event.
Emergency services	Damage to hospitals, ambulances, police and fire stations, and emergency vehicles.	Response efforts will prioritize critical services, such as transportation, hospitals, emergency shelters, communications, utilities, and damage control (prevention of the spread of disease, fire, or hazardous materials). Hospitals and ambulances will be overwhelmed and challenged to handle lifesaving needs.

Myths about Disasters

A few myths exist about disaster response behavior. There is a belief that there will be mass hysteria and that people will panic and loot abandoned buildings as seen in Hollywood movies. This myth is just that—a myth. In reality, most people will respond in a calm and organized manner and help one another.

Another misconception is that people will be too dazed and disoriented to function appropriately. This is also incorrect. Experience has shown that disaster survivors will guide themselves and others to safety. In many disasters, residents have self-organized themselves to assist the injured, assess damage, and initiate search and rescue before emergency responders have even arrived. In many cases, the quick and effective reaction of the community has saved lives.

Another common myth is that most people will evacuate to public shelters. History has shown that survivors understand these places are overcrowded and instead try to stay with a friend or relative. Shelters will not be the primary or preferred option for most people.

Despite the intensity of disasters and the associated emotional trauma, in general people behave very constructively in these situations. Keep an eye on recent disasters that have happened and been reported in the news, and take note of how people helped one another. This is the norm, not the exception.

Common Hazards

Diverse geography and growing climate change make every region susceptible to numerous hazards—and potentially a combination of several types of disasters at once. This guidebook focuses on practical action, so I do not delve extensively into each type of disaster. Nonetheless, I have provided a brief description of a few major disasters to give you some background; this is particularly important to read if you live in a region prone to any of the hazards described below.

Extreme Weather

We are experiencing more extreme weather patterns than ever before, mostly attributed to climate change. Here are some examples:

- severe storms, high winds, or hail
- thunderstorms and lightening
- heavy rain or flooding
- hurricanes or tornadoes
- ice storms or heavy snowfall
- extreme heat or drought

These weather patterns can last for hours or weeks, causing injury, illness, accidents, property damage, and social and economic disruption. Regardless of where you live, everyone should be prepared for the possibility of extreme weather.

Wildfires

Wildfires have occurred naturally for thousands of years and are essential to keep our forests healthy. But the prevalence of catastrophic wildfires is rising, and climate change may be partly to blame. Drought and lightning storms, coupled with human carelessness and population growth in closer proximity to forests, have increased the risk of wildfires.

Wildfires can greatly disrupt a community, forcing evacuations, causing extensive damage, and destroying property, all with significant social and economic impacts.

Aside from the immediate health risks associated with the fire and smoke, aftermath hazards include toxins in the air, water, and soil from burned buildings. If you live near forested or grassland areas, understand how wildfires might affect you and follow the 7 Steps to prepare your family, home, property, and business to reduce the risk of damage.

Floods

Floods are one of the most common disasters. They develop when the ground becomes oversaturated with water and is no longer able to absorb excess water. They occur frequently and can happen at any time of year for a number of reasons:

- heavy rains
- rain over snowpack during the fall and winter
- ice jams
- reservoir releases or dam failures
- coastal flooding from high tides, storm surges, and waves
- heavy rain and rapid snow melt in the spring (known as "freshet"), typically between April and July

Earthquakes

Earthquakes are caused by a sudden release of energy that has accumulated over time along a fault (a weak link in the Earth's crust). The release of energy causes a sudden displacement of rock along the fault, which sends out waves of vibration causing the earth to shake and damaging the built environment. Earthquakes are one of the most destructive natural hazards: They are unpredictable, can occur at any time, and can happen suddenly and without warning. They can also provoke a tsunami, landslides, and fires. Earthquakes cause injury, death, major structural damage, and massive disruption.

Many factors determine the magnitude and intensity of the shaking: the location and depth of the earthquake's epicenter (its origin); the size, type, and direction of the fault line; and the direction of ground movement. The magnitude of the earthquake is measured on the Richter scale, which quantifies the amount of energy released by the earthquake itself from 0 to 9. For this reason, there is only one magnitude number for any given event. For example, a measure of 3.5–5.4 indicates that the tremors are often felt, but rarely cause damage. A magnitude of 7.0–7.9 or higher indicates a major earthquake that can potentially cause serious damage. If you live in or close to an area that is at risk of earthquakes, preparing your home and family can drastically reduce the possibility of injury and damage.

Tsunamis

A tsunami is a series of long, surge-like waves most often caused by strong underwater earthquakes but also by undersea events such as volcano eruptions or landslides. A tsunami can also be activated when a strong local earthquake causes a landslide. The topography of the coastline and the ocean floor will influence the size of the tsunami waves. Likewise, the location and intensity of the earthquake will affect the height, speed, and timing of the tsunami. Tsunami waves can have speeds of up to 560 mph (900 km/h) at the epicenter and travel several kilometres inland. They can be very dangerous and destructive.

Waves may come any time after an earthquake, but most will arrive between thirty-five minutes and two hours after the initial shock. The water levels may rise many feet depending on the location. A popular misconception reinforced by Hollywood movies is that tsunamis consist of one huge wave. In reality, there are several waves separated by minutes, or even up to an hour. Generally, the first wave is preceded by a recession of the sea, and the second and third waves are the largest. While tsunamis are rare, if you live along the coast, it is best to be prepared for such an event.

Communicable Diseases and Pandemics

In this age of mass population, massive cities, and international travel, communicable diseases are a real and common risk. In fact, some experts claim that the next major disaster is more likely to be a pandemic than a weather-related event. Pandemics are a growing global health concern and cause significant illness, death, and social disruption. Additionally, given that pandemics are generally not localized or contained to one area, they can spread very quickly and affect people in every industry, which can seriously disrupt all aspects of a community.

Pandemics and communicable disease outbreaks can be considered a disaster on their own; however, they can also be a side effect of other types of disasters. Causes for communicable disease following a disaster include contaminated food and water, poor hygiene, limited sanitation, and a compromised health care system. As well, people are more vulnerable to disease when they are fatigued from trauma. Emergency shelters may be overcrowded, increasing the risk of contracting communicable diseases. The effects of a pandemic or outbreak are often compounded by the reduced availability and quality of both health care and public health prevention services that may follow any type of disaster. Understanding the dynamics of communicable diseases and how they can spread is an important part of planning for such an outbreak.

Hygiene and Sanitation

Following a disaster, you are likely to face some hygiene and sanitation challenges. There may be no electricity, clean water, or garbage collection, and the sewage system may be damaged. Preparing for this situation will help prevent illness and protect your family's health. Disasters will compromise sanitation and personal hygiene for many reasons:

- Damage to homes may limit access to your normal toilet facilities and to clean water to wash hands, bathe, and launder clothes. Access to personal hygiene supplies, such as toilet paper, towels, soap, and a clean change of clothes may also be restricted.

- Your home may contain extensive debris and dust. Roads may be damaged, preventing transportation of goods and limiting essential sanitation services.

- Electricity and garbage collection services may not be available for extended periods.

- Clean water may not be available due to damaged pipes or contaminated water sources.

Following a disaster, it is particularly important to be mindful of the increased risk of disease that can arise from unsanitary conditions. People may be fatigued, emotional, hungry, and thirsty, and some may be living in overcrowded shelters lacking privacy for personal hygiene activities. These conditions increase vulnerability to illness and the risk of a communicable disease outbreak such as influenza, meningitis, or tetanus.

Children, pregnant women, the elderly, and people with chronic health conditions are particularly at risk of contracting communicable diseases. For these reasons, is it important to protect yourself and your family from illness, and to limit the transmission of disease to vulnerable people who may not be able to recover.

Practicing proper hygiene and sanitation (including diligent hand washing), keeping your vaccines up to date, and handling food and water safely can all help protect against the spread of disease. Read on for more information on these and other practices. I've included a hygiene and sanitation checklist at the end of this section.

Basic Practices

Vaccines

Given the unsanitary conditions in both overcrowded shelters and private homes during a disaster, and the increased risk of contracting communicable diseases such as influenza, meningitis, and tetanus, it's best to ensure your vaccines are up to date. Visit a travel or vaccine clinic annually, preferably in the fall to receive the seasonal influenza vaccine, which is generally available after the summer.

Food and Water Safety

Following disasters, dirt and germs from the environment can get into your food and water. To ensure your food and water are safe to consume, store your food in a clean place and ensure diligent safe food practices. Drink only potable water. If you must disinfect the water, follow the disinfection instructions as noted in Collecting and Disinfecting Water, starting on page 89. That section also suggests safe food and water practices.

Handwashing

The unsanitary conditions following disasters increase the of risk illness. Proper handwashing is essential and is the single most effective way to protect against communicable diseases. Avoid touching your mouth, eyes, ears, and genitals with dirty hands. Wash your hands frequently.

If water is available:

- Rinse hands first.
- Wash for twenty seconds with soap and water.
- Dry with disposable paper towels.

If water is not available:

- Use a hand sanitizer with at least sixty percent alcohol.
- Apply enough sanitizer to fill a cupped hand.
- Rub sanitizer all over, including your thumbs and the backs of your hands (thumbs are often forgotten).
- Let your hands air dry.

It is important to know when to wash your hands. Do so under all of the following circumstances:

- When your hands are soiled.
- Before and after going to the washroom or changing diapers.
- After blowing your nose.
- Before putting anything in your mouth.
- Before and after eating, handling food, or touching raw meat.
- Before disinfecting water.
- After handling garbage, waste, or animals.

Handling Garbage and Other Waste

In the event of a disaster, there may be no garbage collection for a while, so it is important to deal with garbage appropriately. Open garbage releases odors and attracts rodents and insects, and these creatures increase the risk of disease transmission, either by direct contact with people and their belongings, or through their feces and urine, which carry many viruses, bacteria, and parasites. Following a disaster, accumulated garbage in the streets can also negatively affect the morale of the community, reducing motivation for recovery.

Collect your garbage in a safe and tidy manner and encourage your community to do the same. Reduce, reuse, and compost as much as you can, particularly if there is no garbage collection for a long time, as this will greatly reduce the volume of your trash. Keep all trash in tightly closed containers, sheds, or buildings. Another temporary option is to dig a hole, place the bagged garbage inside, and then fill the hole with enough dirt to cover the garbage and prevent odors—about one to two feet (thirty to sixty centimeters) of dirt. Later, when garbage collection resumes, you can dispose of the garbage as you normally would.

Human Waste

Proper disposal of human waste (feces) may be a challenge after a disaster. Disasters may damage sewer systems, pipes may be broken, and sewage treatment plants may not function because there is no electricity to operate the equipment or personnel to monitor it.

Solid waste must be managed in a safe and hygienic manner to avoid health problems. Similar to collecting garbage, human waste must also be collected to prevent the spread of disease, the attraction of rodents and insects (and the accumulation of rodent excrement and urine), and the contamination of wells and other water sources.

If the sewage system is compromised, you may need to set up emergency toilets as per the following guidelines.

Emergency toilets

If the sewage system is intact and damage affects only the water lines, you may not have incoming water but you can still use your toilet by taking the following steps:

- Close the main water valve.
- Using non-potable water (from a pool, hot tub, lake, or rainwater), pour water in the toilet to flush the waste down.

If the sewage system is compromised, do not flush the toilets. Flushing may spill raw sewage into your home. Instead, be prepared to collect and dispose of your waste in a sanitary manner. Ensure you always separate liquid (urine) and solid waste. A camping toilet also works well.

Collect urine in a separate container, such as a large pail, preferably with a tight-fitting lid. Urine is sterile and can be discarded outside in a hole about two feet (sixty centimeters) deep. Pour some dirt on top of the urine to reduce odors and prevent attracting rodents or insects. Mark off the hole in such a way that nobody accidently falls in. If it is comfortable for you, you can also urinate directly outside.

You can use plastic bags in your existing toilet, a bucket, or a container (preferably with a lid and a handle) to collect solid waste.

- Close the main water valve.
- If using the toilet, empty the water first using a small container.
- Line the container (or toilet) with a heavy plastic bag. Apply some disinfectant to reduce odor and destroy bacteria (see disinfectant options below).
- After use, sprinkle more disinfectant on the waste.
- Once the bag is about half full, tie it off loosely to allow air to circulate to help dry out the waste.
- Place the bag in a bin or hole in your yard. Once the sewage system is operating again, dispose of the waste in the toilet.
- Always wash your hands thoroughly after handling the waste.

If the sewage system is out for a while, your community may want to construct field latrines and dispose of the waste properly later. There are several ways to construct field latrines; your local environmental and public health authorities will provide direction for this undertaking.

Options for disinfectants

You can choose from a number of disinfectants. Always use caution with these chemicals and follow package directions.

- Chlorine bleach: Use one part liquid household bleach to ten parts water. Do not use powdered bleach; it is very corrosive.
- Calcium hypochlorite: This chemical is often used in pool maintenance. Follow the directions on the container.
- Portable toilet chemicals: These are available at recreational vehicle supply stores and sometimes hardware stores. Use according to the package instructions.
- Powdered chlorinated lime: Lime will disinfect and also help to dry the waste more quickly. This product is generally available at building supply stores. Be cautious: Lime can be irritating on the skin, so it's best to wear gloves for application.
- Cat litter can also be used to absorb liquids, but most products generally do not disinfect.

Liquid waste

Wastewater from bathing, laundry, and cooking is considered liquid waste and should be drained away from living areas. Liquid waste can attract flies and rodents that transmit disease. Stagnant liquids can become a breeding site for mosquitos and other disease-carriers, increasing the risk of diseases such as filariasis and Japanese encephalitis.

In many disaster situations, people tend to bathe and wash their clothes near lakes and rivers. This practice is not recommended, as it can easily contaminate camps and water sources. Bathing and washing clothes should be conducted in an area designed for such activities. These areas should be screened for privacy, have a supply of water, and be sloped to drain water so as not to contaminate any food or water sources.

Hygiene and Sanitation Checklist

- ☐ Be mindful of the increased risk of disease and unsanitary conditions following a disaster.
- ☐ Wash your hands frequently and thoroughly.
- ☐ Start composting if you're not doing so already.
- ☐ Make sure your trash container has a tight-fitting lid. Purchase one if yours does not.
- ☐ Make sure you have a shovel to dig a hole. Even a small folding spade will do the trick and is not too expensive.
- ☐ Keep two buckets with tight-fitting lids and handles, large plastic bags, gloves, and a disinfectant (chlorine bleach, calcium hypochlorite, or powdered lime) on hand.
- ☐ Avoid having stagnant water pool around your home.
- ☐ Keep your vaccines up to date.
- ☐ Collect and disinfect water safely.
- ☐ Store and handle food safely.

Collecting Water from Your Hot Water Tank

Providing it has not been damaged, your hot water tank is an excellent source of up to forty gallons (150 liters) of water that is safe to drink once it has been properly drained and disinfected. For this reason, it is important to secure your tank to prevent it from falling during a disaster, particularly an earthquake (see Step 5 for instructions). To properly empty your tank you will need a clean garden hose with a female fitting end, a flat head screwdriver, and gloves to protect your hands in case the water heater is hot.

- Shut off the electricity to the heating element. If it is powered by gas, the pilot light should go out.

- If there is a possibility that the city's water is contaminated, close the valve for the incoming water to prevent it from entering your tank. You can reopen it once the water is safe for consumption.

- Attach one end of the garden hose to the bottom of the drain unit and place the other end in a clean container. Open the valve with a screwdriver.

- If no water comes out, there may be a negative vacuum. Open the valve on top to allow air in, then close it when the water starts draining.

- Once the container is filled, close the drain valve.

- Filter and disinfect the water before consumption (see the next section).

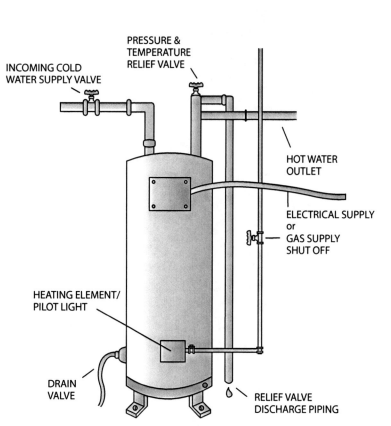

INCOMING COLD
WATER SUPPLY VALVE

PRESSURE &
TEMPERATURE
RELIEF VALVE

HOT WATER
OUTLET

ELECTRICAL SUPPLY
or
GAS SUPPLY
SHUT OFF

HEATING ELEMENT/
PILOT LIGHT

DRAIN
VALVE

RELIEF VALVE
DISCHARGE PIPING

Collecting and Disinfecting Water

Following a disaster, you will require an abundant amount of water for drinking, personal hygiene, food preparation, and cooking. Ideally, you would stockpile enough water to support your family's needs until the city water has been restored. However, the city water may take longer to be restored than expected, and you might need to collect and disinfect water from alternate sources.

Disinfection Methods

Untreated water may contain pathogens (protozoa, bacteria, and viruses) that cause diseases such as gastrointestinal illness, vomiting, cramps, salmonella, shigella, E. coli, norovirus, or hepatitis. To protect your health and the health of your family, any water of questionable quality should be properly treated before being used for drinking, brushing your teeth, cooking, or personal hygiene.

There are a few methods for removing pathogens: boiling, chemical disinfection, ultraviolet light, solar radiation, sedimentation, and filtering. Boiling is the most effective method to remove pathogens. If boiling is not available to you, a combination of the other methods is recommended. Salt and heavy metals from ocean water, on the other hand, require distillation to make the water safe to drink.

Boiling

Boling water is the single most effective method to destroy pathogens and is the recommended method to disinfect any untreated water. However, boiling requires fuel, and you will need to plan accordingly. To effectively destroy pathogens, heat the water to a rolling boil for at least one minute. Cool, then drink.

Chemical Disinfection

If boiling is not available to you, water can also be disinfected chemically. Chemical options include unscented household bleach that contains 5.25 to 6.0 percent sodium hypochlorite, or chlorine tablets or powder. Iodine has been used in the past but is not often used anymore, and it is difficult to find in stores.

If using bleach, use one quarter teaspoon per gallon of water (or one eighth teaspoon per two liters) and let stand for thirty minutes to ensure pathogens are destroyed. The water should have a slight scent or taste of bleach. If it doesn't, let it stand for another fifteen minutes. Be aware that bleach loses half its effectiveness in six months and has a shelf life of sixteen months. On each new bottle you purchase, write the date you purchased it and its expiry date. Note the expiry date on your inventory checklist and on a family calendar or in your smartphone. Be sure to replace your bleach when it gets too old to be effective.

Chlorine tablets or powder are convenient to have in your grab-and-go bags and are available at most sporting goods and camping stores. Follow the product's instructions for dosing and contact time.

Ultraviolet Light

Ultraviolet light (UV) is an effective way to disinfect small amounts of clear water. However, if the water is cloudy, the germs may be shielded from the light by the particles. UV light can reduce pathogens but it is not as effective as boiling or chemical disinfection. Although UV lights are commercially available, they are not easy to find as they are not as effective or as popular as other methods. Follow the product's instructions to ensure proper disinfection.

Solar Radiation

In an emergency situation, sunlight can be used to disinfect water. The water must be in a clear container placed flat on a warm surface, such as on pavement or a roof. You can also place the container on a reflective surface (such as aluminum foil) to better attract the light. Let the bottle stand for six hours in bright sunlight. If the water is cloudy, this technique will not be effective.

Filtration Methods

While boiling destroys pathogens, including giardia and cryptosporidium, other methods (chemical disinfection, UV, and solar radiation) will not destroy these two pathogens. The water will require filtration. You should filter the water to remove these pathogens before disinfecting with chemicals, ultraviolet light, or solar radiation. If boiling the water, there is no need to filter it.

There are several filtration options available at most camping or outfitter stores. Commercial filters can range from 0.3 to over forty microns, and are either absolute or nominal. An absolute filter will be more consistent at filtering, whereas a nominal filter allows twenty to thirty percent of small particles to pass through. Giardia and cryptosporidium are very small and can only be filtered through filters that do not allow particles larger than one micron to pass through.

If you are in a pinch, there are also several homemade filtration options, such as layers of paper towel, coffee filters, or clean cloths. Homemade filters are only somewhat effective, but they are better than not filtering at all. An absolute one-micron filter has been proven to be the most effective filtration method.

Another filtration method is reverse osmosis, which can remove bacteria, viruses, and salt. Military ships use a sophisticated reverse osmosis system to desalinate seawater while at sea. Personal use systems are commercially available, but they are pricey and hard to find.

Sedimentation

Sedimentation is the process of allowing dirt to fall to the bottom of a container over time and then transferring the water to a clean container, leaving the sediment behind. If the water has a lot of sediment, you can apply this process up to three times, each time allowing the sediment to settle, then transferring the clean water.

Distillation

Salt water is not safe to drink; it will make you ill. You can treat saltwater by distilling it, which will remove the salt and most chemicals. This is an easy process, but it takes time. It involves boiling the water and collecting the vapor, which will condense back into liquid without the salt and impurities. There are several commercially available distillation products. The following is an easy method to distill water with what you already have in your home:

- Fill a pot halfway with the water to be distilled.
- Using a lid that fits on the pot and has a handle, tie a cup onto the lid so that when the lid is inverted, the handle points into the cup.
- Turn the lid upside down on the pot. The cup should not be touching the water.
- Boil the water for twenty minutes, or until it is completely evaporated.
- As the water boils, vapor will collect on the lid and drip down into the cup.
- The water collected from the vapor is safe to drink.

Collecting and Treating Water from Alternate Sources

If you are collecting water from an alternate source that has visible debris, you will want to choose a combination of sedimentation, filtration, and disinfection to make it safe to drink. The methods you choose depend on how dirty the source is and what you have available to you.

Collect the Water

- Select as clean a water source as possible, preferably away from animal and human waste or sewage runoff.
- Collect the water with clean hands, tools, containers, and equipment.
- Avoid floating material.
- Allow any particles and sediment to settle to the bottom of the container. Gently transfer the cleaner water to a secondary container. Continue this sedimentation process until the water is clear.
- Filter and disinfect the water according to the instructions below.

Determine Next Steps

- If the source is muddy, sedimentation, filtration, and disinfection are all required.
- If the water is cloudy, filter it first, then apply a double dose of disinfectant.
- If the water is clear, only filtration and disinfection are required.
- Once the water is disinfected, ensure safe and clean storage and be careful not to contaminate the water with dirty hands and equipment.

Filter

- Filter the water using an absolute one-micron filter or a filtering pump system available at most camping or outfitter stores. Make sure to follow the manufacturer's instructions.

- You can also filter through layers of paper towel, coffee filters, or clean cloths. However, this method is not as effective as using a one-micron filter.

Disinfect

- Once the water is filtered, disinfect it by boiling or using a chemical.

- If boiling, bring the water to a rolling boil for one full minute. The water is now potable.

- If using a chemical disinfectant, you can use regular unscented household bleach that contains 5.25 to 6.0 percent sodium hypochlorite, or chlorine tablets or powder.

    - If using bleach, add one quarter teaspoon of bleach per gallon of water (one eighth teaspoon per two liters), stir, and let stand for thirty minutes. The water is now potable.

    - If using chlorine tablets or powder, follow the product's instructions for dosing and contact time.

Disaster Mental Wellness

Directly or indirectly, people will inevitably be affected by disasters. The loss of a loved one, a pet, a home, or a business may cause a variety of challenging emotions. However, proper planning can reduce the intensity of this emotional trauma.

Psychosocial Responses to Disasters

If you are a disaster survivor, you will experience a variety of emotions following the traumatic event. You and your loved ones will have different responses, to varying degrees, and at different times. Some feelings are common immediately after the event, while other emotions may come months afterward.

Intense reactions and emotions are a natural part of the grieving process and generally diminish weeks to months after the traumatic event. People who are generally resilient are able to recover well. Proper planning can be very beneficial in helping you to better manage these situations. Below are some common reactions following a disaster, as well as guidelines to prepare your family.

Common Behaviors

Although everyone will respond differently to disasters, studies have shown that people will respond with different stages of emotions with varying transition times. Although this is not an exhaustive list, the following lists highlight some common reactions to disaster situations.

Initial reactions

- Disbelief, shock, denial
- Fear, confusion, anxiety
- Fear for your family's safety
- Disorientation, numbing, apathy
- Excessive need for information
- Difficulty making decisions
- Excessive need to help others (heroism)
- Reluctance to leave home

Subsequent reactions

- Frustration, irritability, worry, anger
- Grief, sadness, depression, apathy, crying
- Suspicion, anxiety
- Increased susceptibility to colds, allergies, and flu
- Guilt over not being able to prevent the disaster
- Feeling overwhelmed or powerless
- Changes in appetite, sleeping, and digestion
- Changes in eating, smoking, or drinking behavior
- Changes in who you talk to or trust, where you travel, and how you spend money
- Physical reactions including pain, skin rashes, and stomach or bowel irregularity

- Frequent unwanted thoughts or images
- Sensitivity to loud noises
- Strained interpersonal relationships
- Isolation from family and friends
- Domestic violence

Children's responses

Children will respond to a disaster with a variety of emotions. The intensity and duration of disturbing emotions depends on a few factors: age, whether they had direct exposure to the event, whether a friend or family member died, whether they have experienced previous trauma, and how long they are away from their normal routine (home, friends, toys, playground).

Children will react to a disaster based on the response of their parents and other adults in their lives. If parents respond calmly, the children will manage better. If children are exposed to repeated images of the event (e.g., on television), their reactions will likely be intensified. After children have recovered, reminders of the event may trigger more emotions. If your children's symptoms do not subside, you may consider seeking a professional to help them cope with their emotions following a disaster.

While every child will respond differently, the following are a few common reactions:

- Clinginess, crying, screaming
- Sadness, fear
- Nightmares
- Fantasies that the disaster never happened
- Refusal to return to school
- Withdrawal, immobility

Reducing the Emotional Effects of a Disaster

Being prepared is the best way to reduce the effects of a disaster—have a plan, make your grab-and-go bags, and know how respond and recover. Discuss the plan with your family and neighbors. The following are additional tips to reduce the emotional effects of disasters.

Before a Disaster

- Be prepared for a disaster. Develop an emergency plan and get supplies ready—maintain and discuss your plan regularly. Talking about the plan will decrease fears and give people confidence that they can effectively manage a disaster situation.

- Learn about how an emergency or disaster can affect your family's feelings and behavior.

- Talk with your children about emergency preparedness and their role during the event.

- Know your neighbors and maintain connections.

During a Disaster

- Keep informed about the situation and share the facts with your children—as well as the plan to keep them safe.

- Calm any fears that your children may have about being killed, injured, or left alone, or that another disaster will happen. Stay with them and hug them often. The contact will make them feel safe and protected.

- Take care of your immediate needs—eat, rest, and drink water.

- Appreciate your strengths and acknowledge your weaknesses. Ask for help as needed.

After a Disaster

- Be patient; know that this is a process and that the situation will improve in time.

- Take care of yourself—eat well, drink plenty of water, exercise, and get enough rest.

- Talk about your feelings. Ask for help from your friends and family.

- Understand that others are also experiencing the trauma in their own way—be tolerant of their behaviors.

- Participate in memorials to facilitate closure.

- Focus on daily activities. When it feels comfortable, return to your normal routine.

- If symptoms do not improve over time (in a month or so), seek professional help from a counselor, pastor, clergy member, or other trusted advisor.

- Talk with your children about their feelings about the emergency. Encourage them to draw and paint the events. This will help you understand their perspective and what they are going through.

- Reassure them often that they are safe.

- Hold them. Touch provides reassurance that they are safe, protected, and not alone.

- Relax somewhat from the normal rules and praise responsible behavior.

- Communicate with teachers and caregivers to help all of you understand better how your children are managing the trauma.

Special Needs

Some people have unique needs and require special equipment or extra assistance during and after disasters. If you or a member of your family has special needs, additional planning may be required.

Planning for Specific Requirements

Disasters can happen very quickly. You may have little warning before needing to evacuate, or you may be confined to your home without water, electricity, communications, or help from emergency responders. Everyone's ability to cope and specific requirements are unique. That's why it's important to evaluate your needs and prepare for all kinds of emergencies accordingly. For example, if you require medication, it's a good idea to store extra medication in your emergency kit. Consider each family member's needs when creating an emergency plan and gathering supplies.

In addition to the suggestions previously noted for a communication and reunion plan, shelter and evacuation plan, and emergency supplies, the following are some further guidelines to consider when planning for those with special needs:

- Identify your specific requirements for all kinds of emergencies.

- Discuss your emergency plan with your physician to ensure you have included everything you might need.

- Discuss your needs and your plan with your family, friends, neighbors, and landlord, and perhaps your employer.

- Know your neighbors and ask them if they will have any special requirements in the event of a disaster.

- Become familiar with any disaster notification plans in your building and community.

- Get involved with emergency planning in your community to ensure your needs are included in the plan.

- Plan to have someone check in on you after a disaster. Keep in mind that telephones may not be working. You may consider developing a personal support network of three people who are within walking distance who can check on you during an emergency.

- Be aware of financial exploitation scams. Unfortunately, there may be some people who will try to take advantage of your vulnerability, needs, and urgency to recover after a disaster. Avoid high-pressure salespeople, and never disclose personal financial information to sales or service companies without a legitimate written contract.

Supplies

- Store your emergency supplies in a kit that you can carry, such as a backpack or a suitcase with wheels. Label your kit and special equipment with your name and address.

- Make a list of your allergies and medications. Ask a pharmacist to print out a list of your prescriptions to make this task easier. The list should include the name and dose of each medication, what condition it is meant to treat, and your physician's name and address. This list may come in handy if someone takes care of you or if you forget some details in the stress of a disaster.

- Store extra medication in your grab-and-go bag. Inquire with your pharmacist if you can have a sixty- to ninety-day emergency supply, in case pharmacies are not accessible in the event of a disaster. Tell the pharmacist that it is for your emergency kit and ask for the latest expiration date.

- If you need glasses, keep an extra pair in your grab-and-go bag.

- If you require regular service from a health provider or personal care agency, inquire if they have a plan to provide service following an emergency.

- Consider a medical alert system that will allow you to call for help if you need assistance.

- Keep a whistle and small flashlight in handy locations around the house. This will help alert others to your location in an emergency situation. The international signal for help is three short blasts.

- Place a battery-operated night light in each room.

- If you have a service dog, keep extra food and water in your kit and at home. Practice evacuating with your dog. Be aware that shelters typically do not accept pets, unless they are service animals.

Special Equipment or Devices

- Include extra batteries in your kit for any devices or equipment you may need. Also, consider using generators or backup power sources.

- Secure oxygen tanks and life support equipment so that nothing falls over in a disaster. Have a plan for backup power for the system.

- If you require equipment to assist with walking, ensure it is handy, or have backup devices in another part of the home and with a friend.

- If you use a wheelchair, consider several possible wheelchair accessible exits out of the building and practice emergency and fire drills. Stock your chair with medicines, sanitary aids, a flashlight, and a horn to signal for help. Consider including a tire patch kit, seal-in-air product, and spare inner tubes as well.

Seniors with Special Needs

It is especially important to prepare in advance to assist seniors with special needs. The following are a few considerations to be given to seniors.

- Find out if, and where, special assistance is available from the community.

- Visually impaired persons are typically very familiar with their immediate surroundings. However, in an emergency situation, objects may have been displaced and people can become disoriented. Inform them about the emergency and guide them until they are in a safe environment.

- People who are hearing impaired may not hear the emergency warnings. Let them know what has happened by writing it out or using sign language, and guide them until they are safe. Encourage them to make special arrangements ahead of time so that they can receive warnings.

- People with mobility challenges may need special arrangements and equipment.

Specific Health Conditions

If you or a loved one has a specific health condition, consider wearing a medic alert tag. If there are any allergies, specific dietary needs or sensitivities, or special medications, include a supply of appropriate food or medication in your grab-and-go bag.

Hearing impaired

- Consider a battery-operated television for receiving information in a power outage.
- Keep a pencil and pad handy for communicating.
- Plan to have friends help you in an emergency.
- Consider installing a smoke detector and fire alarm system with a strobe light function.

Vision-impaired

- Keep extra walking devices around the home.
- Have extra vision aids as required.
- Familiarize yourself with evacuation routes in advance and consider how you will manage them.

Diabetes

Consider carrying an extra supply of the following:

- Fast- and long-acting insulin, oral agent
- Glucagon injections, dextrose tablets, sugar
- Syringes, needles, insulin pens, sharps container
- Blood glucose-testing kit, lancets, and/or blood glucose and urine ketone strips
- Spare batteries
- Extra food, meal replacement drinks or bars
- A cooler or thermal bag with ice packs for the insulin
- Copies of your prescription
- Contact information for your physician and endocrinologist

Pregnant women

If you are or expect to be pregnant, have an emergency delivery kit handy, which includes the following:

- Non-sterile gloves (three pairs)
- Clean towels to hold and wrap the baby
- Blankets
- A nose syringe for suctioning the baby's mouth
- Plastic bag for the placenta—you may want to show it to the doctor.
- Thread, twine, or shoelaces for tying the umbilical cord
- Fresh sanitary pads
- Sterile cotton swabs
- Skin scissors in the event an episiotomy is required

Disaster Insurance

Insurance is an important step in preparing for emergencies and disasters both large and small. Although it is a common belief that insurance covers injury, illness, and damage to homes and may cover fire, wind, and theft, it often excludes earthquakes, floods, and tsunamis.

It is important to review your policies in detail to ensure they adequately cover your home and vehicle for the hazards in your area. Know about the benefits and limitations of a standard policy and of possible additional insurance for earthquakes, floods, and sewer backups.

Consider purchasing additional insurance for disasters to defray the costs of damage or loss. This could mean the difference between a smooth recovery and significant financial loss. Such additional insurance can also provide some peace of mind and reduce the stress associated with a disaster.

General Insurance Tips

The following are some general tips on disaster insurance and the benefits, limitations, and options that may be available to you. Keep in mind that each insurance company provides different coverage, so it is best to check with your provider for the details.

Living Expenses

In certain circumstances, homeowners who are unable to return to their home as a result of insurable damage are entitled to compensation for additional living expenses. However, coverage may be limited. Some companies may offer living expenses for only two to four weeks and may include a maximum claimable amount. Ensure the expenses to meet your family's needs are covered following a disaster.

Disaster Restoration

In the event of a disaster, your insurance company will assess the damage and appoint a restoration company. Repairs will then begin, but depending on the magnitude of the disaster and the resources available, repairs can take several months, if not longer. During a large event, restoration companies from across the country will help out. Nonetheless, it may take time.

As a result of the delay in restoration, you may be tempted to start the repairs yourself. If you do, be aware that your insurance company may not endorse your work. Furthermore, if there are any future issues with your repairs, they may not be covered. However, it is generally safe to take up any efforts to mitigate the damage, such as removing water or debris and cleaning up.

Selecting a Contractor for Repairs

Depending on the damage to your home, you may be able to repair many things yourself. However, you may need to hire a contractor for some of the more technically involved repairs and clean-up, such as fire and flood remediation, and structural repairs. If a contractor is not assigned to you by your insurance company and you need to hire one, be aware of scams.

Following a disaster, there will likely be a number of people taking advantage of your vulnerability and urgency. They may claim to be qualified and experienced to conduct the work. However, here are a few things to consider when hiring contractors:

- ✔ Are they licenced and in good standing with their industry and professional association?
- ✔ Are they covered by workers' compensation insurance?
- ✔ Do they have general liability insurance?
- ✔ Are they best suited for the work, with adequate experience?
- ✔ Are they willing to sign a legitimate contract with reasonable terms and conditions?
- ✔ Are they asking for a large deposit? (Never pay for the entire job in advance.)

Although you may be anxious to start repairs, hiring a reliable contractor will potentially save you money and frustration later.

Disaster Finance Assistance

Some countries, states, or provinces may provide some financial assistance for uninsurable property. Recommendations for such support are typically made by emergency management authorities. You may want to contact them in advance to understand the eligibility criteria and what this assistance might cover.

Insurance Checklist

- ❑ Review your house, vehicle, life, and business insurance policies, and know about the benefits and limitations of your standard policy.
- ❑ Ensure you are covered for damage caused directly or indirectly by the hazards in your area, such as fire, flood, earthquake, and sewage backup.
- ❑ Ensure you have an up-to-date will.
- ❑ Decide if additional insurance is required to ensure coverage for your home and vehicle for the hazards in your area.
- ❑ Find out if your insurance policy includes replacement value and what out-of-pocket expenses are covered if you must evacuate your home.
- ❑ Find out what benefits and services your insurer will provide after an insured loss, including additional living expenses.
- ❑ Make an inventory of your home and belongings, preferably with receipts and digital images. Digitally secure them on an Internet site and store copies with an out-of-area friend or family member.
- ❑ Continue regular upkeep of your home to reduce the risk of damage: check your sump pump monthly, clear debris from gutters, anchor any fuel tanks, raise your electrical components above your home's projected flood elevation, etc. See the General House Preparation Guidelines starting on page 35 for more details.
- ❑ Consider retrofitting your home to reduce the risk of damage.

❑ If disaster finance assistance is available to you, have the documents and information handy.

❑ If you are a tenant, consider purchasing earthquake insurance to cover the contents of your home. Have a discussion with your landlord to discuss his or her plan in the event of an earthquake or a flood. Be prepared for the possibility that your rental home will not be available following a disaster and that you will need to plan to live elsewhere, perhaps staying temporarily with family or friends.

The decision to purchase earthquake insurance is often based on your current financial situation, acceptable financial risk in the event of major damage, the value of the home, and the importance of safe shelter and a smooth recovery for your family. This decision can be challenging, making it especially important to clearly understand the benefits and limitations of additional insurance.

Shopping List

The following is a shopping list of items you may need to purchase for your emergency supplies. It does not include items that you are likely to have at home already, such as spare clothing, toiletry items, pens, and paper. Adjust quantities based on how many kits you make and how many family members you have. Download a printable version of the annual maintenance checklist from my website, at www.authorkimfournier.com/resources/ (password *flashlight*).

For a detailed description of the emergency kits and where to store them, see Step 4 on page 19. Details on preparing your house can be found in Step 5, starting on page 35.

Miscellaneous

- ❑ Grab-and-go bags: a larger one for the house; smaller ones for the car, work, and school (see page 20 for details on suggested type of bag)
- ❑ Storage containers or bag for home supplies (see details in Disaster Supplies for the House, page 25)
- ❑ File folders or envelopes (preferably waterproof) to store copies of important documents
- ❑ A corded phone (does not require electricity, unlike cordless phones)
- ❑ Alternative cell phone charger, such as a solar, hand-cranked, or vehicle charger
- ❑ Surge protector bars to protect sensitive equipment from power surges
- ❑ Earplugs
- ❑ Sunglasses
- ❑ Toothbrush, toothpaste, dental floss
- ❑ Flip flops for public showers
- ❑ Powdered laundry soap
- ❑ Small blanket
- ❑ Several candles, various sizes
- ❑ Extra batteries for flashlights and any other devices you may need
- ❑ Pocket-sized game or deck of cards
- ❑ Disposable waterproof camera
- ❑ Household chlorine bleach

Outfitter and Camping Supplies

- ❑ Multi-purpose soap
- ❑ Bug spray
- ❑ Chlorine tablets to disinfect water

- ❑ Waterproof whistle
- ❑ Several flashlights or headlamps, small and large (consider non-battery-powered options like solar or wind-up)
- ❑ Portable battery-powered or wind-up radio
- ❑ Multipurpose tool
- ❑ Waterproof matches, lighter
- ❑ Bike chain, lock and key, tire repair kit, and air pump
- ❑ Two-way radios (walkie-talkies)
- ❑ Poncho or raincoat
- ❑ Emergency blanket (silver, heat-reflecting)
- ❑ Sleeping bag
- ❑ Tent (or plastic sheeting)
- ❑ Cots, air mattresses
- ❑ Compact survival book

Hardware

- ❑ Fireproof safety box
- ❑ Several fire extinguishers
- ❑ Ladder
- ❑ Fire escape ladders
- ❑ Garden hose
- ❑ N95 mask
- ❑ Double-sided tape
- ❑ Work gloves, safety goggles
- ❑ Rope, zap straps, duct tape
- ❑ Snow shovel, sand shovel
- ❑ Broom
- ❑ Plastic sheeting or vapor barrier
- ❑ Tarps
- ❑ Sheets of plywood and wood pieces for repair
- ❑ Latches for cabinets
- ❑ Safety film for windows
- ❑ L brackets
- ❑ Eye hooks and wire
- ❑ Recreational vehicle antifreeze

- ❑ Blanket or insulating material to protect pipes and valves from freezing
- ❑ Tools: knife, hammer, multi-screwdriver, two wrenches, axe, hacksaw, crowbar, various screws, nails
- ❑ Consider: generator, emergency toilet, sump pump

First Aid Kit

- ❑ Non-latex disposable gloves
- ❑ Alcohol-based hand sanitizer
- ❑ Bandages, gauze, elastic bandage
- ❑ Medical tape
- ❑ Face masks
- ❑ Slings and splints
- ❑ Cotton tipped swabs, disinfectant
- ❑ Sanitary napkins (these are multi-purpose, very absorbent, clean, and excellent for large wounds)
- ❑ Multi-purpose scissors, tweezers
- ❑ CPR mask
- ❑ Thermo blanket
- ❑ Medication: anti-nausea, anti-diarrhea, antihistamine, analgesics or painkillers, antibiotic ointment
- ❑ Pocket-sized first aid manual (available at most safety, camping, or outfitter stores)

Water

- ❑ At least three to seven days' supply of water; aim for three weeks' worth. For details on the quantity of water you will need, and how to prepare and store your water, see page 26.

Food

- ❑ At least three to seven days' supply of food; aim for three weeks' worth. For details on the quantity of food you will need, see page 28.
- ❑ Canned beans, lentils, legumes
- ❑ Marinated or canned vegetables
- ❑ Canned fruit, soup, pasta, chili
- ❑ Quick-cooking dried noodles, rice, oatmeal
- ❑ Protein bars, dehydrated foods, fruit, nuts, crackers
- ❑ Tea, coffee, powdered milk, spices, peanut butter
- ❑ Dehydrated meals or ready-to-eat foods, such as those purchased at camping stores
- ❑ Candies, gum

Cooking

- ❏ Basic cooking supplies: manual can opener, ladle, spoons, forks, knives, bowls, plates, cups (collapsible cups and bowls are handy)
- ❏ An alternative cooking stove, such as a candle warmer, fondue pot, camping stove, or charcoal grill
- ❏ Spare fuel
- ❏ Collapsible dishwashing container (can also be used for personal hygiene)

Pets

- ❏ Leash or harness, tied on the outside of the grab-and-go bag for easy access
- ❏ Pet food
- ❏ Water and a drinking or eating container
- ❏ Two weeks' supply of pet medication
- ❏ Pet booties
- ❏ Poop bags, paper towels, hand sanitizer
- ❏ Litter box, litter, and a scoop
- ❏ Cleaning supplies, such as soap and bleach
- ❏ Consider a crate if needed

Acknowledgements

Thanks to Nicole Beneteau and Anna Wärje for their thoughtful pre-editing. Their initial feedback was a big help in shaping the manuscript.

I am indebted to Cara Robbins for her amazing graphic design, layout and illustrations. She made my ideas come alive on paper!

A very special thanks to Karen Crosby of Editarians for her insightful, visionary, passionate, and skillful editing.

Last but certainly not least, my gratitude goes out to my family, friends, and closest colleagues for their support.

About the Author

Kim Fournier, CD, MA, is a disaster management consultant specializing in practical and real results for today and sustainable solutions for tomorrow. Well known for her practical, innovative, and compassionate approach, Kim integrates needs, capacity, training, practice, and research to cultivate resilient families, businesses, communities, and government organizations.

Kim's unique expertise has been honed throughout her dynamic thirty-plus year career. She integrates her experience in community outreach and health, public education, survival, disaster management, preparedness, and response, as well as international, interdisciplinary, and humanitarian operations. She served for twenty-five years with the Canadian Forces, for which she has been awarded numerous medals honoring her missions, accomplishments, and dedication.

Kim is passionate about helping families and communities cope with disaster, contributing her diverse skills and knowledge to global disaster risk reduction and resilience. She feels privileged to be part of a global disaster risk reduction team that strives to improve disaster resilience and self-reliance for this generation and the ones to come.

Products and Services

I understand that disaster resilience requires ongoing commitment and education. As such, I offer additional products and services to make your family, community, business, or organization more resilient. Visit me online at www.kimfournier.com for more information about these products and services.

- Emergency preparedness manuals tailored for your community, agency, or organization
- Emergency preparedness presentations and workshops
- Training packages tailored to your situation
- Checklists, guides, and reference sheets for your home or office
- And more …

Stay connected at www.kimfournier.com and on my Facebook page (facebook.com/kimfournierconsulting/) to join the discussion and keep up to date about emergency preparedness news, research, lessons learned, and the latest tips and equipment. See you online!

CPSIA information can be obtained
at www.ICGtesting.com
Printed in the USA
LVOW09s1842050218
565348LV00009B/640/P